# Tribology

## (Friction, Wear, Lubrication and Bearing)

# TRIBOLOGY

## (FRICTION, WEAR, LUBRICATION AND BEARING)

*By*

**Krishan Kant Sharma**

*Astt. Professor, Department of Mechanical Engineering*

*Galaxy Global Group of Institutions*

*Ambala, Haryana*

UNIVERSITY SCIENCE PRESS

(An Imprint of Laxmi Publications Pvt. Ltd.)

**BANGALORE • CHENNAI • COCHIN • GUWAHATI • HYDERABAD**
**JALANDHAR • KOLKATA • LUCKNOW • MUMBAI • RANCHI**
**NEW DELHI • BOSTON, USA**

*Published by :*
UNIVERSITY SCIENCE PRESS
(*An Imprint of Laxmi Publications Pvt. Ltd.*)
113, Golden House, Darya Ganj,
New Delhi-110 002

*Phone* : 011-43 53 25 00
*Fax* : 011-43 53 25 28

www.laxmipublications.com
info@laxmipublications.com

*First Edition* : 2013

**OFFICES**

| | | | |
|---|---|---|---|
| ✆ **Bangalore** | 080-26 75 69 30 | ✆ **Chennai** | 044-24 34 47 26 |
| ✆ **Cochin** | 0484-237 70 04, 405 13 03 | ✆ **Guwahati** | 0361-251 38 81 |
| ✆ **Hyderabad** | 040-24 65 23 33 | ✆ **Jalandhar** | 0181-222 12 72 |
| ✆ **Kolkata** | 033-22 27 43 84 | ✆ **Lucknow** | 0522-220 99 16 |
| ✆ **Mumbai** | 022-24 91 54 15 | ✆ **Ranchi** | 0651-220 44 64 |

UTR–9692–125–TRIBOLOGY–SHA C—

*Typeset at* : Sukuvisa Enterprises, New Delhi. *Printed at* : Repro Knowledgecast Limited, Thane

# Preface

I am glad to present the book entitled, 'Tribology (friction, wear, lubrication and bearing)' for the engineering students of mechanical, civil, automobile and aeronautical. The course contents have been planned in such a way that the general requirements of all engineers, researchers and students are taken care of. Because there were problems in recommending any single textbook which containing all the units according to University Syllabus. Existing textbooks were either too specialized or too literal. Many books provided exhaustive reviews of friction, lubrication, bearing and wear data while others provided detailed description of the lubrication and wear problems occurring in machinery.

In this book various concepts explained in the simplest possible terms with supporting illustrations and simple diagrams to understand the working of that part/topic. The book is written in a simple and easy-to-follow language, so that even an average students can easily grasp the subject by self-study.

At the end of chapters many questions and unsolved problems are given to solve. Extra diagrams and last year's University Question Papers have also been included where it was found that the readability of the original text could be improved. The subject matter of the book is also relevant to mechanical and materials engineering, applied chemistry and physics courses.

I express my appreciation and gratefulness to my parents, publisher, colleagues, friend and students who encouraged me to write this book.

**—Author**

# CONTENTS


*Preface* (*v*)

**1. Tribology** **1—8**

1.1 Introduction 1

1.2 Meaning of Tribology 2

1.3 Learning Tribology 6

1.4 The Several Disciplines in the Field of Tribology 7

**2. Wear** **9—39**

2.1 Introduction 9

2.2 Terminology in Wear 11

2.3 Tribology of Polymers 33

2.4 Tribology of Polymer Composites 34

2.5 Oxidative Wear 36

2.6 Methods of Wear Measurement 38

**3. Friction** **40—62**

3.1 Introduction 40

3.2 Theories of Friction 43

3.3 Coefficient of Friction (COF) 47

3.4 Angle of Friction 49

3.5 Stick-Slip Phenomenon 51

3.6 Friction Measurement Methods 57

3.7 Uses of Friction 59

3.8 Causes of Friction 61

**4. Lubrication** **63—91**

4.1 Introduction 63

4.2 Purpose/Functions of Lubricants 64

4.3 Properties of Lubricants 67

4.4 Lubricant Additives 72

4.5 Functions/Objective of Lubrication 76

4.6 Advantages of Solid Lubricants 80

4.7 Hydrodynamic or Full Fluid Film Lubrication 81

4.8 Boundary Lubrication 83

4.9 Elastohydrodynamic Lubrication 84

4.10 Hydrostatic Lubrication 84
4.11 Gear 85

**5. Bearings** **92–112**
5.1 Introduction 92
5.2 Design of Bearings 92
5.3 Journal Bearings 94
5.4 Reynolds Equation for Squeeze Films 98
5.5 Porous Bearings 99
5.6 Hydrostatic Bearings 103
5.7 Gas Lubricated Bearings 105
5.8 Thrust Bearings 106
5.9 Rolling-Contact Bearings 108
5.10 Forces and Stresses of Rolling-Contact Bearings 109
5.11 Bearing Materials 111

**6. Surface Topography** **113–125**
6.1 Introduction to Tribological Systems 113
6.2 Surface Topography 115
6.3 Surface Examination Techniques 119

**References** **126–127**
**Examination Papers (Previous Year's)** **128–132**
**Index** **133–134**

Chapter 1

# TRIBOLOGY

## 1.1 INTRODUCTION

**Tribology,** the concept which focuses on friction, wear and lubrication of interacting surfaces in relative motion. Tribology is based on that when two bodies moving, sliding on each other then friction occurs Fig. 1.1. Tribology in a traditional form has been in existence a well studied examples of how early civilizations developed in bearings and low friction surfaces. The study of tribology the basic laws of friction, such as the proportionality between normal force and limiting friction force. Understanding of friction and wear in the surface of the bodies, to explain the underlying mechanisms. The relatively low priority of tribology concept that nobody really bothered what would happen when motion between the spheres and surfaces was in a downward direction. Tribology is the field of science which studies the improved friction control. The work of Reynolds initiated countless research efforts aimed at parts of improving the interaction between two contacting surfaces. As a result journal bearings are now designed to high level of sophistication. Wear and the fundamentals effect of friction are far more complex problems, the experimental investigation based on advanced instrumentation scanning electron microscopy and atomic forces microscopy. Tribology is according to the Fig. 1.1 shows is the study of the wear, friction and lubrication.

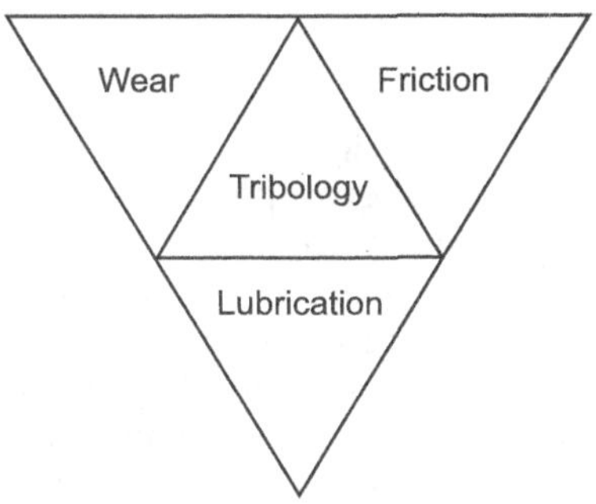


Fig. 1.1

In the field of Tribology affects our lives to a much greater degree than is commonly realized field of science, most of the knowledge being gained after the Second World War. In comparison many basic engineering subjects, *e.g.*, material science, thermodynamics, mechanics and plasticity, are relatively old and well established. Tribology is an imperfect subject to some controversy which has impeded the diffusion. We can say that Tribology is the study of friction, wear, and lubrication.

## 1.2 MEANING OF TRIBOLOGY

The meaning of Tribology based upon the principal causes of wear, friction and lubrication. **'Tribology'** is derived from the Greek word **'tribos'** meaning rubbing or sliding surfaces. On the two surfaces when one surface slides over the another surface. The word 'tribology' has gained gradual acceptance, is relatively new, its meaning is still unclear community and numorous comparisons with tribes or tribolites tend to persist. Wear is the major cause of material wastage and loss of mechanical various stages of performance and any reduction in wear can result in considerable savings. Friction is a principal cause of wear and energy dissipation. Considerable savings can be made by improved friction control surfaces one third of the world's energy resources is needed to overcome friction in one form or another. Lubrication is an effective means of controlling wear, function of lubricant and reducing friction. Tribology which applies an operational analysis to overcomes problems of great economic significance such as reliability, maintenance of technical equipment ranging from household appliances. The majority of problems accounted in the field of tribological. Figure 1.2 shows the study of the wear friction and lubrication in various field of engineering.

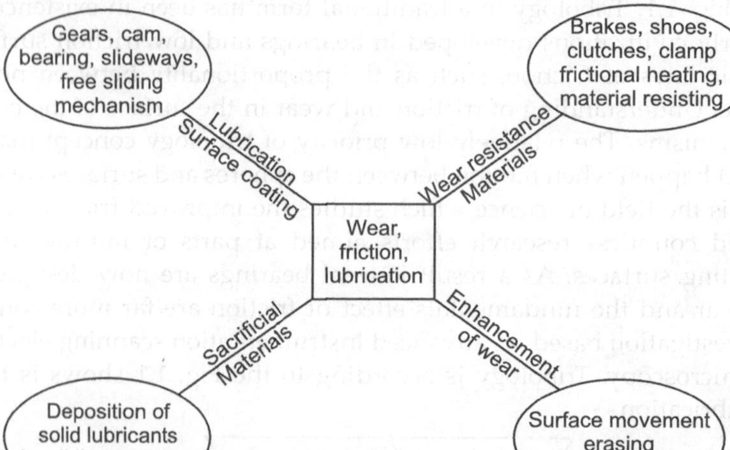


**Fig. 1.2**

In our Daily lives there are so many examples like as our human body also contains interacting surfaces, *e.g.*, human joints, which are subjected to lubrication and wear. Deposition of solid lubricants by sliding the lubrication of human joints is still far from fully understood. Tribology affects our lives much greater degree than is commonly realized. For example, the deliberate control of friction and wear was first promoted, human beings and naturally animals were instinctively modifying friction and wear as it affected their own bodies. Tribology is the contact mechanism of the lubricant Additives, construction, mechanical and chemistry and Environmental Engineering. It is common knowledge that the engine becomes noisey and vibration as a response to stress or damage. It has only recently been discovered that damage on the parts of the engines due to the wear and friction and less uses of the lubricants between both the surfaces.

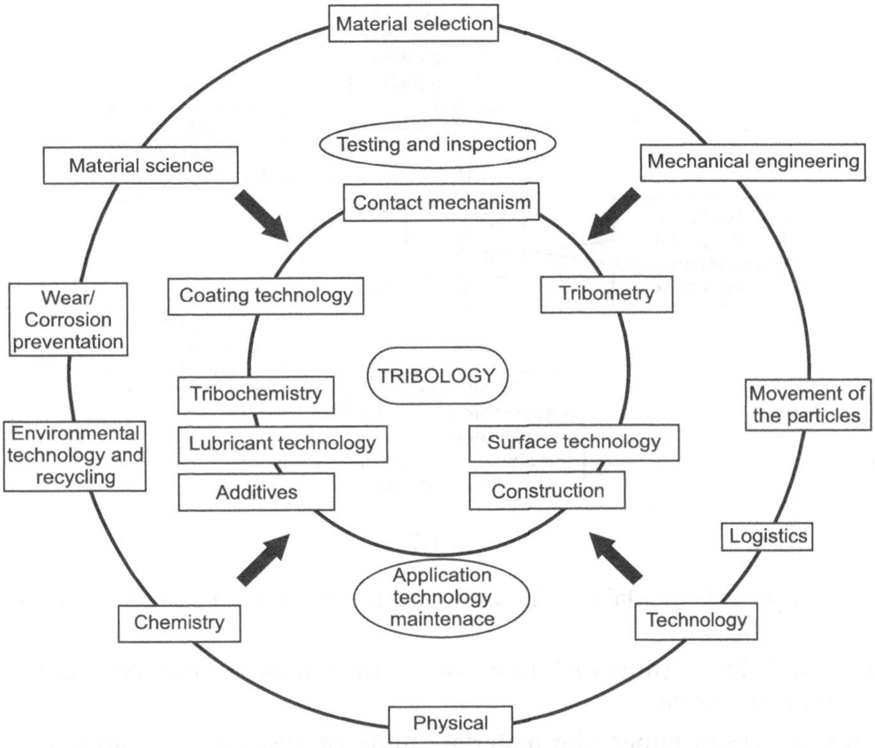


**Fig. 1.3:** Interacting surfaces

In Tribology why 'the interacting surfaces in relative motion', (which essentially means moving, rolling, sliding, normal approach or separation of surfaces), are important to our economy? How and why they affect our standard of living? Surface interaction of particles dictates or controls the functioning of practically every device developed by man in nature. Everything that nature makes wears out, almost always as a result of relative motion between surfaces. An analysis of machine breakdowns, parts missalign show that in the majority of cases failures and stoppages are associated with interacting moving parts such as rollers, gears, bearings, couplings, sealings, cams, clutches, etc. As shown in Fig. 1.4 the Tribocycle is studied the manufacturing systems, design the part, new technology and the selection of the material and surface technologies.

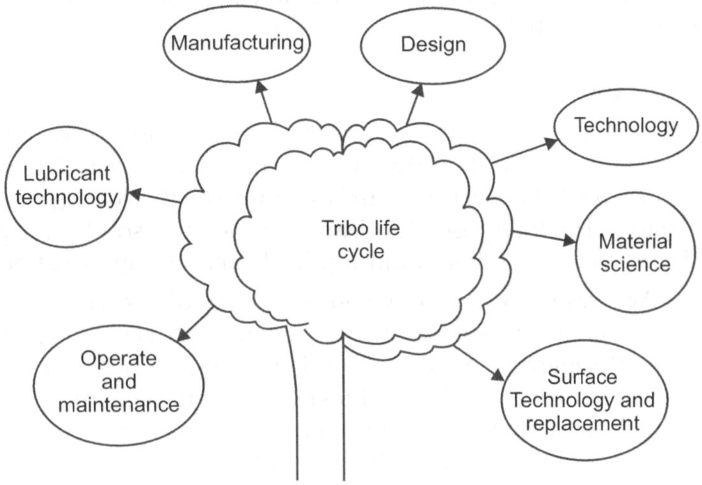


**Fig. 1.4:** Tribo life cycle

To study of the best method of wear, friction, lubrication so that we reduces the power consumption and design out maintenance.

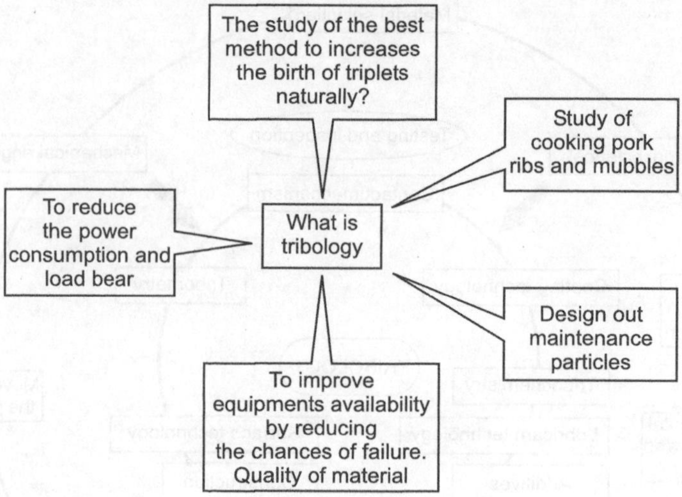


**Fig. 1.5**

To improve equipments availability by reducing the chances of failure, tribology comprises the study of:

- the characteristics of films of intervening surface material between contacting bodies. Design of maintenance.
- the consequences of either film material failure or absence of a film which are usually manifested by severe friction and wear so that improving equipments' ability.

Between surfaces the film formation any pair of sliding objects is a natural phenomenon which can occur with or without human intervention. Film formation might be the fundamental mechanism preventing over extremely high shear rates at the interface between two rigid sliding objects or particles.

### Lubrication

**Lubrication** is the process, or technique employed to reduce or minimize wear of one or both surfaces in close proximity, and moving relative to each another, by interposing a substance called *LUBRICANT*, between the surfaces to carry the load (pressure generated) between the opposing surfaces. The interposed lubricant film can be a solid, (*e.g.*, graphite, $MoS_2$) a solid/liquid lubricant dispersion, a liquid, a liquid-liquid dispersion (greases) or exceptionally a gas.

According to Fig. 1.6, when the two bodies surfaces contacting and rubbing then the friction and wear occurs so avoid wear on the surface we generally use lubricants. These layers of material separate contacting solid surfaces, bodies and are usually very thin and often difficult to observe. In general, the thicknesses of these films range from **1–100 [μm]**, although thinner or thicker films. Knowledge of layers that is related to enhancing or diagnosing the effectiveness of these films in preventing damage, film in solid contacts is commonly known as **'lubrication'.** To avoid wear on the surfaces generally uses the lubricants which known as the lubrication between the surfaces. For example, gaseous film is suitable for low contact stress while solid films are usually applied to slow sliding speed contacts. Detailed analysis of gaseous or liquid films is usually termed **'hydrodynamic lubrication'** while lubrication by solids is termed

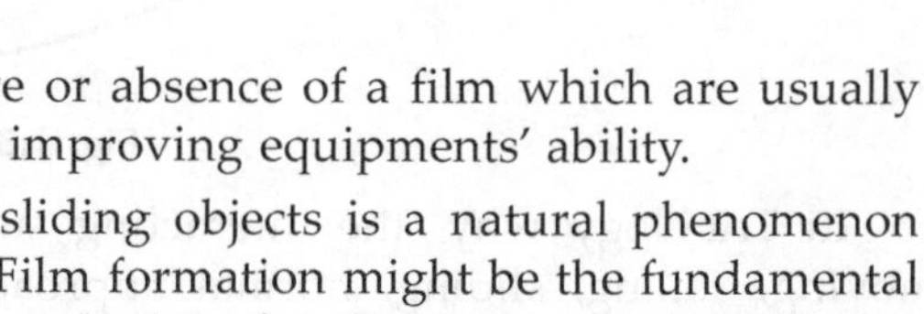


**Fig. 1.6**

**'solid lubrication'**. In the Fig. 1.7 shown the lubricant between the material (A) and material (B) when the bonded elastomers are forced by the applied load. Lubrication is the layer which is filling the gap between the two particles A and B to reduces the friction and wear.

According to Fig. 1.7 there are two surfaces and the gap between them, when load applied and force is also there so that friction and wear occurs to avoid wear a thin film layer is obtained known as lubrication.

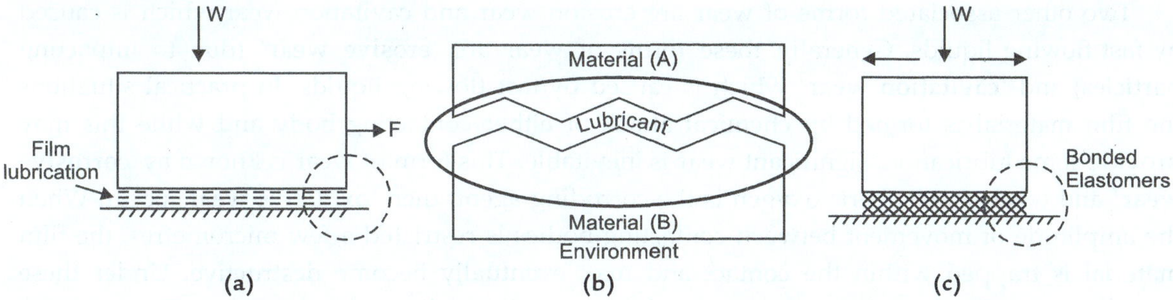


**Fig. 1.7**

In the absence of film, the only reliable means of ensuring relative movement between surfaces is to maintain, by external force fields, a small distance of separation between the opposing surfaces. When the Block moves on the surfaces and the load and force W is applied then wear occurs. So lubricants provided for better performances. A form of lubrication that operates by the same principle, *i.e.*, forcible separation of the contacting bodies involving an external energy source, during movement is **'hydrostatic lubrication'** where liquid or gaseous lubricant is forced into the space between contacting bodies. Liquid lubrication is a technological nuisance since filters, pumps and cooling equipments are required to maintain the performance of the lubricant over a period of time. To avoid the noisey and vibration between both the surfaces we generally provided the lubricants between them. Liquid lubricants are the loss of load carrying capacity at high temperature and degradation in service. The performance of the lubricant depends on its composition and its physical and chemical characteristics, of the surfaces when processes occurs on the surfaces of the material. Prediction methods for liquid or gaseous films involve at the elementary level of boundary, hydrodynamic, hydrostatic and elastohydrodynamic lubrication. For more sophisticated analyses **'computational methods'** must be used.

### Wear

**Wear** is the erosion of material from a solid surface by the action of another surface. It is related to surface interactions and more specifically the removal of material from a surface as a result of mechanical action. The need for mechanical action, in the form of contact due to relative motion, is an important distinction between mechanical wear and other processes with similar outcomes. The consequence of film failure is severe wear. Wear in these circumstances is the result of adhesion between contacting bodies and is termed **'adhesive wear'**. Intervening films are partially effective then milder forms of wear occur and these are often initiated by fatigue processes due to repetitive stresses under either sliding or rolling. These milder forms of wear can therefore be termed **'fatigue wear'**. The film material consists of hard particles or merely flows against one body to another without providing support then a form of wear, which sometimes can be very rapid, known as **'abrasive wear'** occurs.

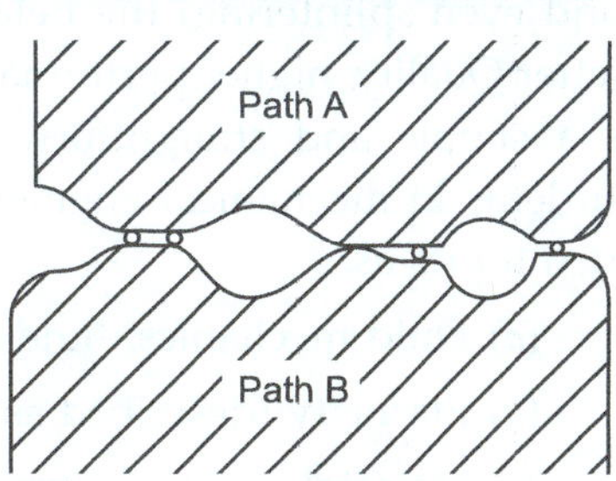


**Fig. 1.8**

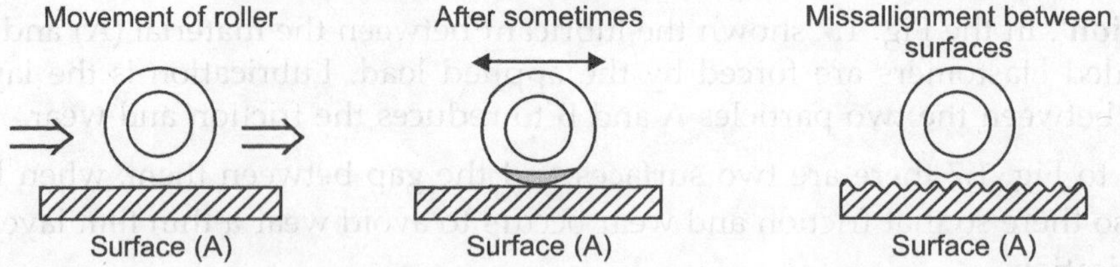


Two other associated forms of wear are erosion wear and cavitation wear which is caused by fast flowing liquids. Generally these forms of wear are **'erosive wear'** (due to impacting particles) and **'cavitation wear'** which is caused by/fast flowing liquids. In practical situations the film material is formed by chemical attack of either contacting body and while this may provide some lubrication, significant wear is inevitable. This form of wear is known as **'corrosive wear'** and when atmospheric oxygen is the corroding agent, then **'oxidative wear'** occur. When the amplitude of movement between contacting bodies is restricted a few micrometres, the film material is trapped within the contact and may eventually become destructive. Under these conditions **'fretting wear'.** Any interaction between solid bodies will cause wear. Examples are **'impact wear'** caused by impact between two solids, **'melting wear'** occurring when the contact loads and speeds are sufficiently high to allow for the surface layers of the solid surfaces to melt, and **'diffusive wear'** occurring at high interface temperatures. This dependence of wear form on various operating conditions can be summarized in a flow chart.

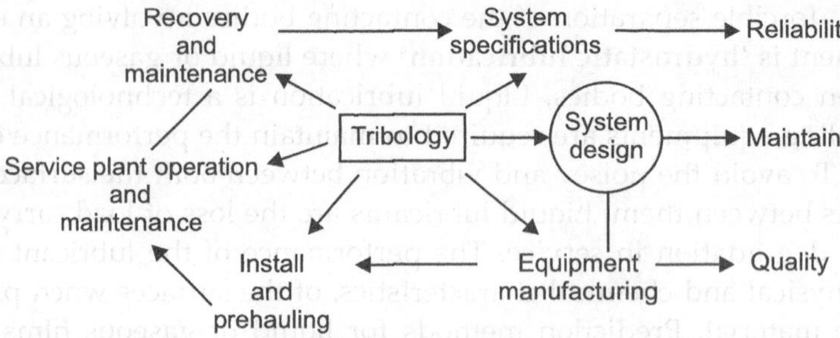


**Fig. 1.9:** Role of Tribology in different espects

## 1.3 LEARNING TRIBOLOGY

Tribology is ultimately an applied knowledge in, many topics like as wear, lubrication, bearings, friction and their classifications. It is not a science by itself although research is done in several different stages to understand the fundamental aspects of tribology. It has the effect of perpetuating (and even splintering) the field along disciplinary lines. Tribology is the system design for the Better Quality, higher performances, Realiability, Installation and system specifications. Tribology is a simple and straightforward art. In academic preparation for designing products, most students in mechanical engineering (the seat of most design instruction) have taken courses in such topics as:

(*a*) Fluid mechanics, fluid machines

(*b*) Elasticity (described as solid mechanics), Ductile, plasticity

(*c*) Materials science (survey of atomic structure and the physics of solids) (*d*) Dynamics (mechanical mostly)

(*e*) Heat transfer

(*f*) Methods of mechanical design, material design.

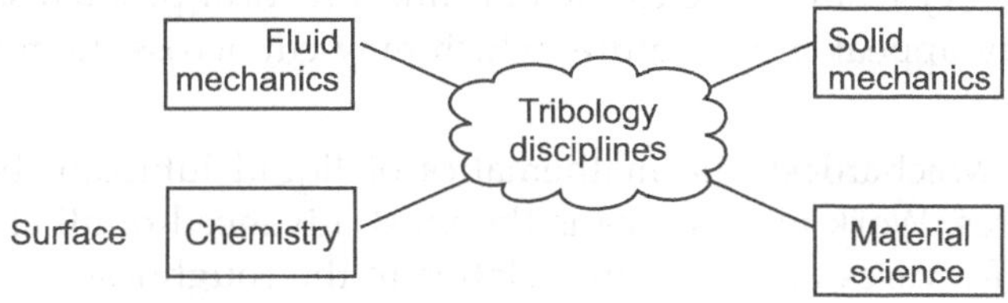


**Fig. 1.10:** Role of Tribology in mechanical engineering

(*g*) Plasticity and elasticity

(*h*) Visco-elasticity

(*i*) Contact mechanics, surface finish

(*j*) The full range of mechanical properties of monolithic materials, composite materials, and layered structures (coatings, etc.), especially fracture toughness, creep, fatigue (elastic and low cycle)

(*k*) Surface chemistry, oxidation, adhesion, adsorption

(*l*) Surface-making processes

(*m*) Statistical surface topographical characterization methods

(*n*) Lubricant chemistry, solid mechanics

(*o*) and several more.

These are useful tools indeed, but hardly enough to solve a wide range of problems in friction and wear.

## 1.4 THE SEVERAL DISCIPLINES IN THE FIELD OF TRIBOLOGY

In the study of the tribological system of material engineering will have different set of tools which are used to justify the concept of Tribology and beware all the mechanics.

**1. Solid Mechanics**: Focus is on the mathematics of contact stresses and surface temperatures due to sliding. friction and wear rates of selected mechanical devices that are based on very simple physical tribological mechanisms. Which are related to the solid mechanism.

**2. Material Science:** The atomic and microscale mechanisms whereby solid surface degradation or alteration occurs during sliding work in this area is usually presented in the form of micrographs, as well as energy spectra for electrons and X-rays from worn surfaces. In the material science studied based on metals and alloys. Firstly, the limit of knowledge in the materials aspects of tribology has not yet been found. Second material scientists (engineers, physicists) rarely have a broad perspective of practical tribology. (Material engineers often prefer to be identified as experts in wear rather than as tribologists.)

**3. Applied Surface Chemistry:** The reactivity between lubricants and solid surfaces. Work in this area progresses largely by orderly chemical alteration of bulk lubricants and testing of the lubricants with bench testers. The major deficiency in this branch of tribology is the paucity of work on the chemistry in the contacting and sliding conjunction region. Work in each of these four areas is very detailed and thorough, and each requires years of academic preparation. There is little likelihood of any person becoming expert in two or three of these areas. The best that can be done is for interdisciplinary teams to be formed around practical problems. Academic programs in general tribology may appear in the future, which may cut across the major disciplines given above.

**4. Auto and Fluid Mechanics:** The mathematics of liquid lubricant behaviour for various shapes of sliding surfaces. Work in this area is the most advanced of all efforts to model events in the sliding interface for cases of thick films relative to the roughness of surfaces. Work is also done on the influence of temperature, solid surface roughness, and fluid rheology on fluid film thickness. The movement of the fluid during the working of the two surfaces and the study of the fluid mechanism during the effect of the lubrication on the body.

Chapter 2

# WEAR

## 2.1 INTRODUCTION

**Wear** is the erosion or damage of material from a solid surface by the mechanical action of another surface. It is related to surface interactions and more specifically the removal of material from a surface. Wear is occurred on the surfaces when friction is there between surfaces.

Wear can also be defined as a process in which interaction of the surfaces of a solid with its working environment results in dimensional loss of the solid, with or without loss of material. Aspects of the working environment which affect wear include loads objectives (such as unidirectional sliding, reciprocating, rolling, and impact loads), speed, temperature, type of counter body (solid, liquid, or gas), and type of contact. When the two bodies at the position of motion the friction takes places on or between the surfaces. This definition also fails to include impact wear, where there is no sliding motion, cavitations, where the counter body is a fluid, and corrosion, where the damage is due to chemical rather than mechanical action.

1. Maintain low contact pressure between surfaces
2. Maintain low sliding speed motion
3. Maintain smooth bearing surfaces
4. Prevent high temperature
5. Use hard materials objectives
6. Insure a low coefficient of friction
7. Use a lubricant.

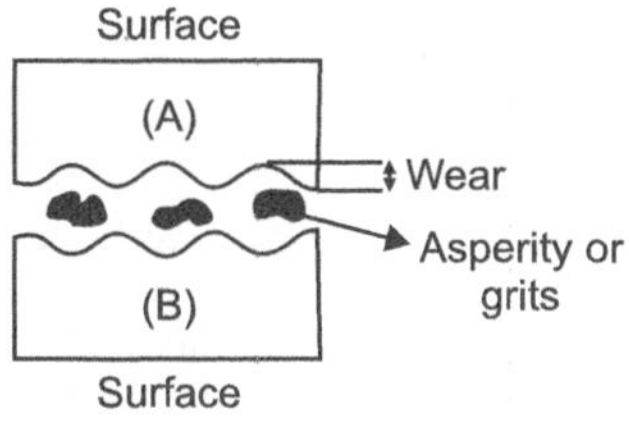


These conditions are not likely, however, to yield a competitive product for maintaining smooth surfaces sliding or prevent high coefficient of friction lubricants. In this chapter a perspective will be provided on what about wear, function and lubrication and the types which occurred on the operation of the friction and movement of the two surfaces. Some machinery eventually fails or becomes uneconomical to operate because of single cause (type of wear), but most mechanical devices succumb to combinations of causes. Wear is based on the motion of the two bodies when the materials are same or the different metals. Predicting the wear life of machinery may perhaps be best understood in terms of the life expectancy of a baby.

**Quantitative Expression for Abrasive Wear:**

$$\Delta L = v.\Delta A = P.\pi r^2$$

$$\therefore \quad dV = r.x.dl = r^2.\tan\theta.dl = \Delta L.\frac{\tan\theta.dl}{\pi P}$$

where

$$\frac{dV}{dl} = \frac{\Delta L.\tan\theta}{\pi P}$$

$$\frac{dV}{dl} = \frac{L\tan\theta}{\pi P}$$

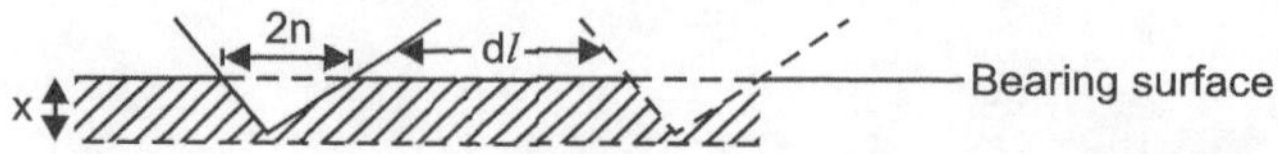


Fig. 2.1(A)

Interaction variables include family history, exposure to diseases and accidents, economic status, personal habits, social context of living, etc. Clearly, life expectancy is not a linear effect of the above variables, and the parallel breaks down in the determination of the endpoint of the process of decline. Quantitative law of wear the subject of wear is based on the shape to describe types, rates, and modes of wear. Wear can be defined as the progressive loss of substance resulting from mechanical interaction between two contacting surfaces. In general these surfaces will be in relative motion, either sliding or rolling, and under load. Wear occurrs because mechanical failure of highly stressed interfacial zones and the failure mode will often be influenced by environmental factors. Surface deterioration can lead to the production of wear particles by a series of events characterized by adhesion and particle transfer mechanisms or by a process of direct particle production akin to machining or, in certain cases, a surface fatigue form of failure and losses of the particles objects movements.

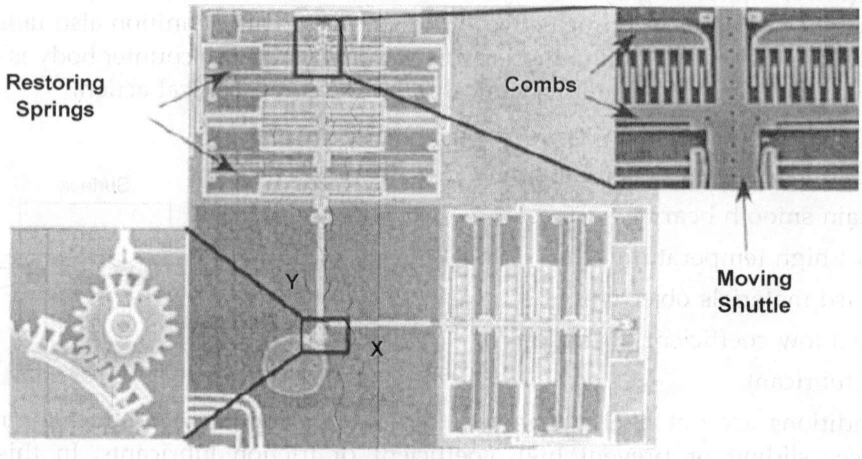


Fig. 2.1(B)

According to the Fig. 2.1 is the gear box mechanism where the gears meshes over each other the restoring springs and combs or shuttle is also in moving condition but after sometimes their teeths of the gear becomes vibratory due to wear in the material and gap occuring between both the gear as shown in Fig. 2.2.

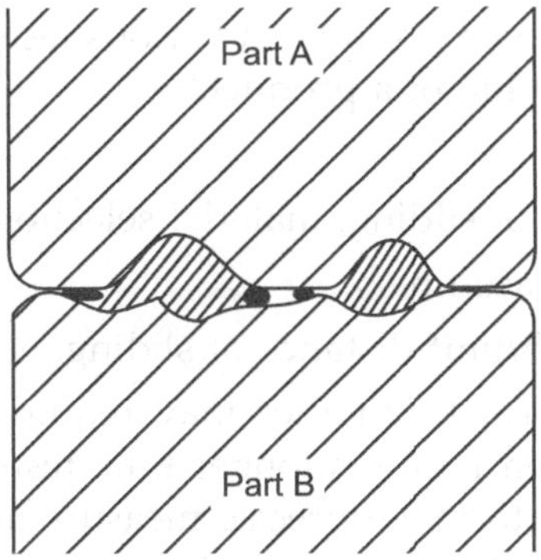


Fig. 2.2

## 2.2 TERMINOLOGY IN WEAR

In the Terminology and measure of the wear has many terms, and several groups in professional societies have worked to provide standard definitions about wear. These efforts are largely attempts to describe complicated sequence of events (chemical, physical, topographical, etc.) The chemical, material science and mechanical topic is studied. Some terms communicate more than others the actual causes of loss (wear) of material from a surface, some are very subjective in nature and the particular wearing process together. Progressing from the more subjective to the more basic on the contact mechanism.

### Mechanisms of Wear

**Wear** is the removal of the material from the surface of a solid body as a result of mechanical action of the counterbody.

**Wear of Engine Bearing** is the removal of the bearing material from its surface caused by sliding friction between the bearing and the crankshaft.

Wear may combine effects of various physical and chemical processes of the friction between two counteracting materials: micro-cutting, micro-ploughing, Plastic deformation, cracking, fracture, welding, melting, chemical interaction, surface roughness and mechanical processes.

***The succession of events whereby atoms, products of chemical conversion, fragments, etc, are induced to leave the system (perhaps after some circulation) and are identified that embodies or immediately suggests solutions.***

These solutions may include choice of materials, choice of lubricants, choice of contact wear volume condition, choice of the manner of operation of the mechanical system, etc.

Wear by abrasion and erosion are forms of wear caused by contact between a particle and solid material. Abrasive wear is the loss of material by the passage of hard particles over a surface. Erosive wear is caused by the impact of particles against a solid surface. Cavitation is caused by the impact of fluid against a surface during the collapse of bubbles. Abrasion and erosion in particular are rapid and severe forms of wear and can result in significant costs if not adequately controlled. The successive form of the atoms or bond which breaks when collides, a particle of liquid can cause erosion. These differences extend to the surface practical consideration of materials selection for wear resistance due to the different microscopic mechanisms of wear occurring in either abrasion, erosion or cavitation adhesion.

### Measurement of Wear and Wear Coefficient

The measurement of material loss to be taken from various tests could take any form that can be used to predict the end of the wear life of a product.

These include:

- Wear volume/unit distance of sliding, malerial selection.
- Mass loss/unit of time, surface finish.
- Change in wear track width/unit distance of sliding, etc.

In the conversion from one form to another may require knowledge of material density. The measurement of mass loss often requires longer time tests than other methods, which has been studied on wear, many ways to make precise measurements of volume loss than to wait for sufficient mass loss to measure. One precise method is to use a surface tracer system and to measure the profile of the wear scar. Several parallel traces can be converted into a volume of material loss. An important caveat in using precise methods is to assure first in a test series that wear is progressing in the same way early in a test as it would late in a test.

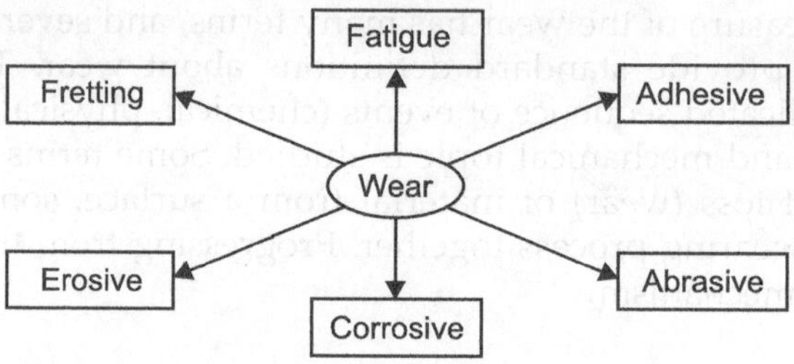


**Fig. 2.3**

PAML MODEL

| Solid contact | Fluid Mechanics | Particle Dynamics | Wear |
|---|---|---|---|
| ↓ | ↓ | ↓ | ↓ |
| Surface contact | Film Thickness Area | Particle Motion Movement | Material Selection and Normal leading |
| Topography $z_{ij} = z(x, y)$ | $h = h\ [z(x, y), \sigma]$ | $\frac{d^2 X_{pi}}{dt^2} = \Sigma f_i$ | Removal Rate $MRR =$ $f$(contact, chemistry, particles), solid particles |
| ↓ | ↓ | ↓ | |
| Elastic foundation $\sigma = \sigma\ (z, f_n, E)$ | $P = P\ [h, n, u]$ | Particle forces $(\Sigma f)$ Applied on surfaces | |

### Types of Wear

Wear may combine effects of various physical and chemical processes proceeding during the friction between two counteracting materials: micro-cutting, micro-ploughing, Plastic Deformation, cracking, fracture, welding, melting, chemical interaction between the surfaces.

*The mechanisms of wear:*

- Abrasive wear
- Adhesive wear
- Fatigue wear
- Corrosive wear
- Erosive wear
- Cavitations Erosion wear.
- Fretting wear

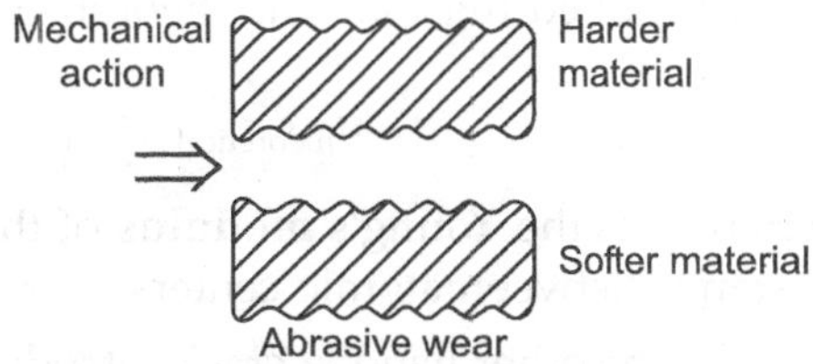


## Abrasive Wear

Abrasive wear occurrs when a harder material is rubbing against a softer material.

- If there are only two rubbing parts involved in the friction process the wear is called **two body wear.**

In this form of wear of the softer material surface is caused by the asperities on the harder surface.

If the wear is caused by a hard particle (grit) trapped between the rubbing surfaces it is called **three body wear.** The particle may be either free or partially embedded into one of the mating surface materials. When the harder surface and the softer surface meshing each other.

Abrasive wear occurrs whenever a solid object is loaded against particles of a material that have equal or greater hardness. A problem is the wear of shovels on earth-moving machinery. If there are only two or more mating parts material, even if the bulk of it is very soft, may cause abrasive wear if hard particles are present. For example, an organic material, such as sugar cane, is associated with abrasive wear of cane cutters and shredders because of the small fraction of silica present in the plant fibres occurred the form of wear.

*Mechanisms of Abrasive Wear*

Abrasive wear by grits or hard asperities resembled cutting by a series of machine tools or a file. However, microscopic examination has revealed that the cutting process is only approximated by the sharpest of grits.

**(*A*) Cutting:** In the abrasive wear mechanism the hard and soft materials.

The basic mechanisms were revealed: a cutting mechanism and a wedge build up mechanism with flake like debris. This latter mechanism, called '**ploughing**', was found to be a less efficient mode of metal removal than '**micro-cutting**'. In the wear has been random plate-like debris were formed by a stylus scratching cast iron. It is probable that in an actual wear situation the effect of cutting alone is relatively small since much more material is lost by a process that has characteristics of both cutting and fatigue.

**(*B*) Fracture:** A **fracture** is the separation of an object or material into two, or more, pieces under the action of stress when the mechanical action occurred on the material surfaces.

*Fracture* is often applied to bones of living creatures, or to crystals or crystalline materials, such as gemstones or metal. Wear in crystalline materials, crystals fracture without the body actually separating into two or more pieces. Depending on the substance which is fractured, a fracture reduces strength (most substances) or inhibits transmission of light (optical crystals).

The crystalline structure of the Brittle material. Fracture can occur by ***clavage*** as the result of **tensile stress** acting normal to crystallographic planes with low bonding. In **amorphous solids,** the lack of a crystalline structure results in a **conchoidal fracture**, with cracks proceeding normal to the applied tension. The theoretical strength of a crystalline material is

$$\sigma_{\text{theoretical}} = \sqrt{\frac{E_\gamma}{r_o}}$$

where: $E$ is the **Young's modulus** of the material, $\gamma$ is the **surface energy,** and $r_o$ is the equilibrium distance between atomic centers.

On the other hand, a crack introduces a stress concentration modeled by $\sigma_{\text{elliptical crack}} = \sigma_{\text{applied}}$ $\left(1+2\sqrt{\frac{a}{\rho}}\right) = 2\sigma_{\text{applied}}\sqrt{\frac{a}{\rho}}$ (For sharp cracks)

where: $\sigma_{\text{applied}}$ is the loading stress, $a$ is half the length of the crack, and $\rho$ is the radius of curvature at the crack tip. Putting these two equations together, we get the young's modulus of the material.

**(C) Fatigue:** The repeated strain caused by grits on surface deforming the area on the surface of a material cause metal fatigue. The young's modulus of the material and the yield properties of the wear the sideways material displacement mechanism shows a transverse section of an abrasion groove. In abrasive wear by repeated sideways displacement of material would a relatively mild or slow form of abrasive wear since repeated deformation is necessary to produce a wear particle.

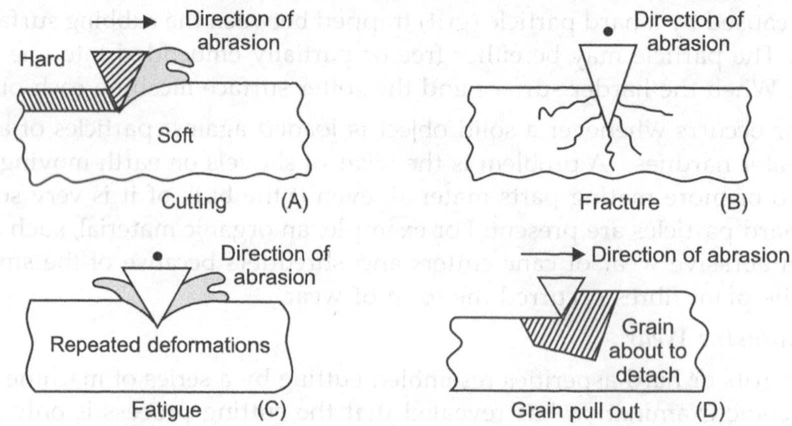


**Fig. 2.4 (A, B, C, D)**

**(D) Grain Pull-Out:** Abrasive grain detachment or pull-out is a relatively rare form of wear which is mainly found in ceramics. This mechanism of wear can become extremely rapid when inter-grain bonding is weak and grain size is large.

***Modes of Abrasive Wear***

The way the grits pass over the worn surface determines the nature of abrasive wear. In the modes or path obtained in the Abrasive wear.

- Two-body and
- Three-body abrasive wear
- Two-body abrasive wear is obtained by the movement action of sand paper on a surface. Hard asperities or rigidly held grits pass over the surface like a cutting tool. In three-body abrasive wear the grits are free to roll as well as slide over the surface, since they are not held rigidly. When the repeated deformation on the surface.

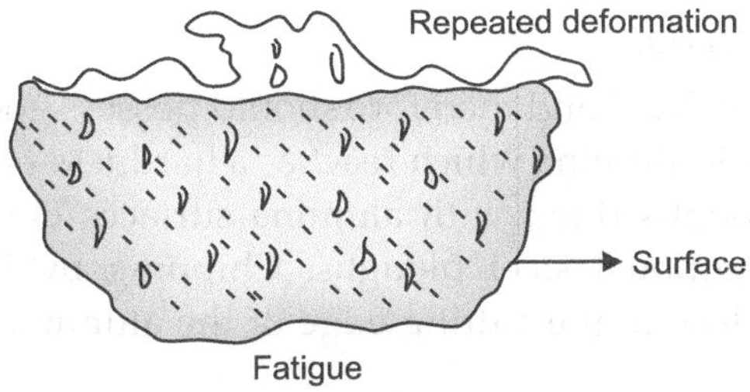


Fig. 2.5

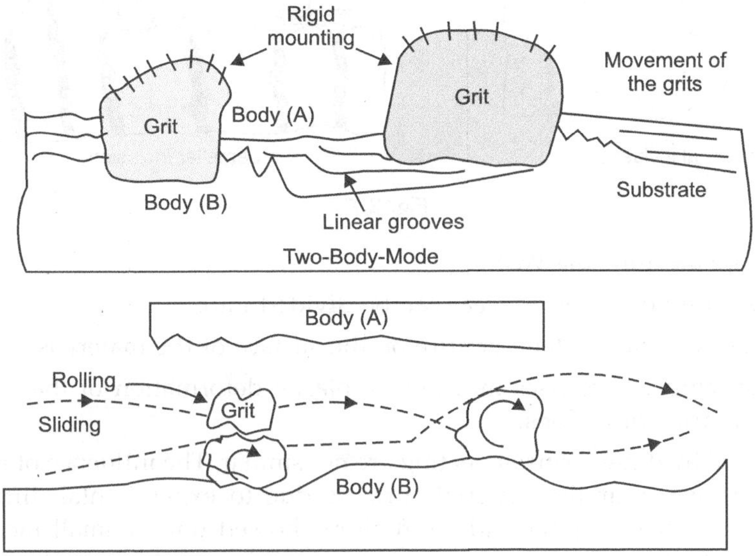


Fig. 2.6 (A)

### *Two and three Bodies of Abrasive Wear*

The abrasive wear resistance material of steels can be considerably enhanced selection of hardness and metallurgical phase. Selection of a steel depends on the hardness of the abrasive particles. For example, if the abrasive is relatively soft, *i.e.*, hardness is less than 1000 VHN, then it is possible to select a steel of hardness that would be greater than **0.8 × hardness of the abrasive** wear, when the two and the three body of Abrasive wear has been studied in this case, abrasive wear resistance is not quite synonymous with hardness. The abrasive wear resistance of a steel to a hard abrasive is determined by the relative proportions of austenite, bainite, martensite, pearlite and ferrite and by the presence of cementite particles in the materials.

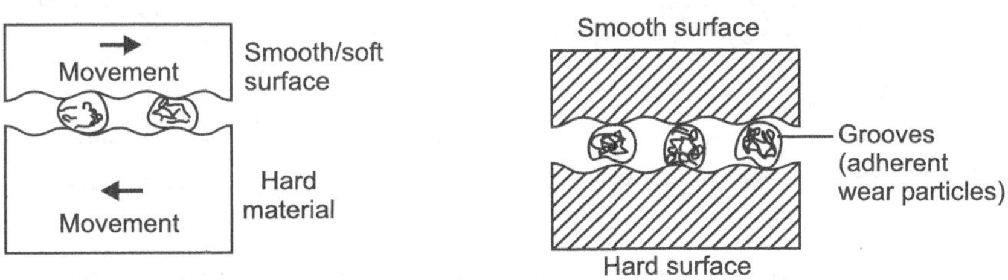


Fig. 2.6 (B)

*Abrasive Wear Resistance Material*

Ceramic materials are extremely hard and therefore should possess good abrasive wear resistance. Material hard ceramic mineral is alumina which reaches a hardness of **9** in the form of corundum. Field tests on agricultural ploughs fitted with alumina surfaces demonstrated greatly reduced wear in comparison to conventional steel ploughs. Chipping by impacting stones, however, resulted in accelerated wear close to the cutting edge of the alumina plough.

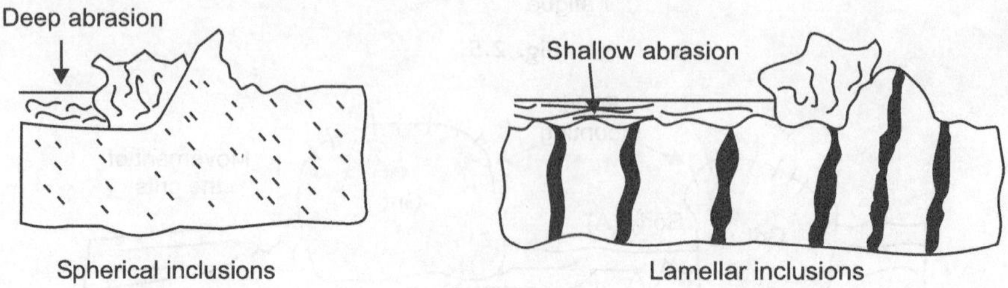


**Fig. 2.7**

*Effect of Temperature on Abrasive Wear*

The effect of temperature on abrasive wear can be divided into:

- The influence of ambient temperature on the surface of the materials.
- The role of temperature rises induced by plastic deformation of the worn material on contact with grits formations.

The effects caused by these forms of heating are not similar. The influence of elevated ambient temperature on abrasive wear the material surfaces due to experimental difficulties. Limited tests of the abrasive wear of copper and aluminium showed only a small increase in wear at temperatures up to 400°C. This trend was recorded in experiments conducted up to temperatures of 2000°C where most metals and metallic carbides and nitrides showed the same proportional decline in hardness of the materials on the basis at a temperature of about '**0.8 × melting point**'.

*Effect of Moisture on Abrasive Wear*

In the abrasive wear temperature and the moisture has a strong influence on abrasive wear rates. Abrasive wear rates increase with moisture content in the atmosphere of the particle when a contrary effect occurrs. Prediction of the moisture effect for any particular case is difficult. The grit may either be weakened by moisture to produce a larger number of new cutting edges, or severe grit weakening and the uniform high temperature and hardness ratio.

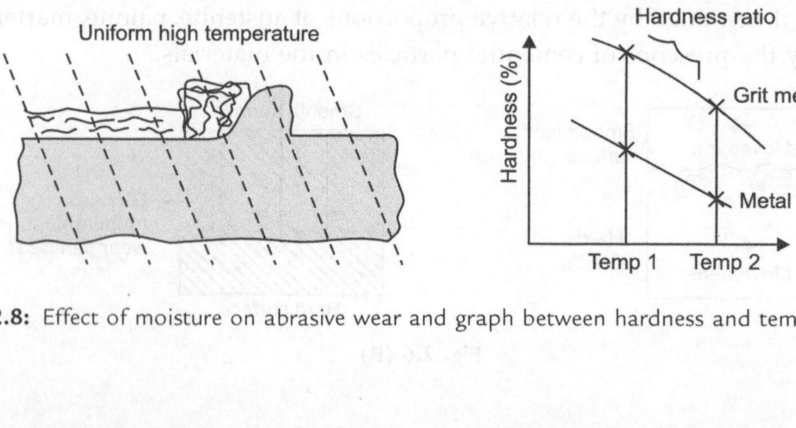


**Fig. 2.8:** Effect of moisture on abrasive wear and graph between hardness and temperature

*Control of Abrasive Wear*

Abrasive wear is the most rapid form of wear and causes the largest costs to industry, several methods have been developed to minimize the losses. The basic method of abrasive wear control mechanism is to raise the hardness of the worn surface until its value is atleast **0.8** of the grit hardness. In the concept of the abrasive wear there are of course complications about prevention of brittleness while raising the hardness which can be overcome only to a certain degree. Abrasive wear is usually suppressed by the application of a hard material or hard coating. Abrasive wear is also controlled by the action of motion of the large temperature rige.

## Adhesion Wear

Adhesion wear is generally be caused by welding rubbing between the opposing asperities on the rubbing surfaces of the counter bodies. When load applied to the contacting asperities is so high that they deform and adhere to each other forming micro-joints. Particles when the motion of the rubbing counter bodies result in rupture of the micro-joints and the breaking different bond of materials. Thus some of the material is transferred by its counter body. This effect is called scuffing or galling. When a considerable areas of the rubbing surfaces are joined during the friction a seizure of one of the bodies by the counter body may occur. Adhesion wear occures when two surface collides each other or very small gap between both the surfaces.

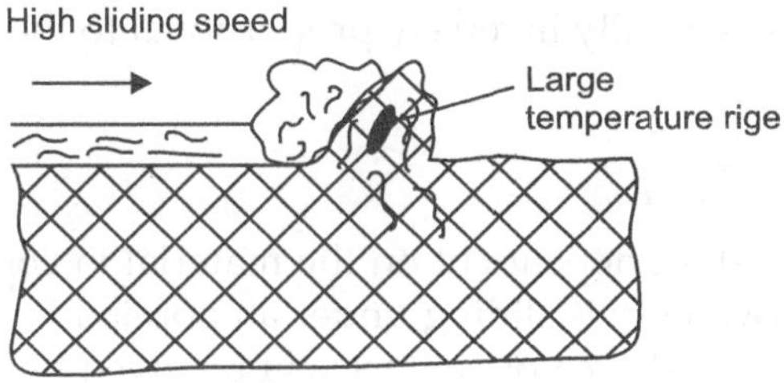


**Fig. 2.9**

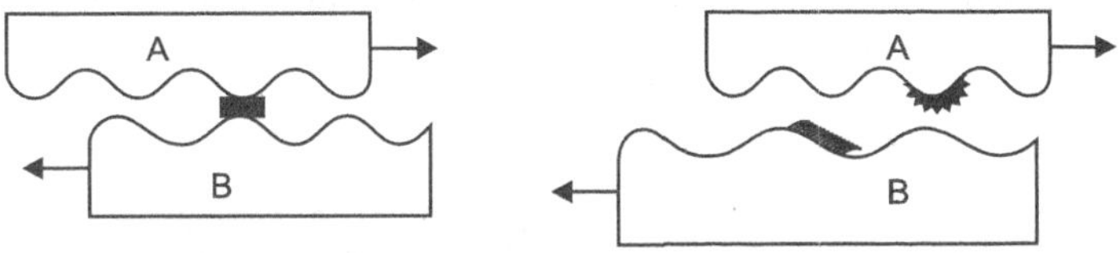


**Fig. 2.10:** Adhesive wear

*The factors decreasing adhesive wear:*

- Lower load on surface.
- Harder rubbing materials, welding joints.
- Contaminated rubbing surfaces.
- Presence of solid lubricants.
- Presence of a lubrication oil, and lubricants additives.
- Anti-wear additives in oil.

Adhesion between two objects particles casually placed together is not observed because intervening particles of contaminant layers of oxygen, water and oil are generally present. In the adhesion wear the two bodies/surfaces which are hard or the soft materials. Adhesion is also reduced with increasing surface roughness or hardness of the contacting bodies. When sliding, rolling and movements between both bodies surfaces. Adhesion and sliding experiments performed under high vacuum totally different tribological behaviour of many common materials from that observed in open air. Metallic surfaces free of oxide films under high vacuum exhibited the most surface particles of dramatic changes and partly for this reason have been widely studied when strong material forces on the weak materials and transfer the bonds on the surfaces.

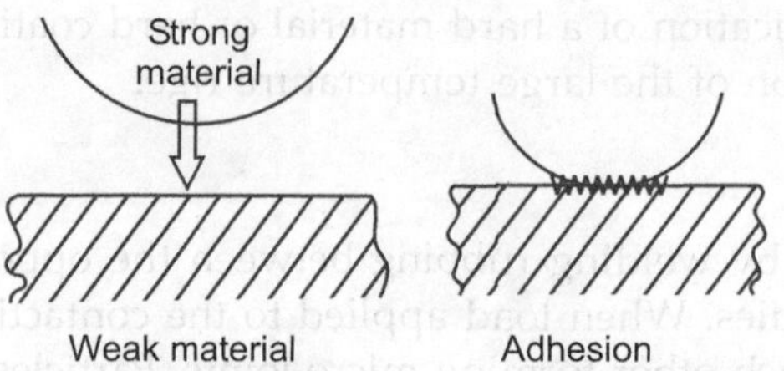


Fig. 2.11

**Quantitative law of adhesive wear:**

(*i*) The amount of wear is generally directly proportional to the load "*L*" On the surface.

(*ii*) The amount of wear is generally proportional to the distance.

(*iii*) The amount of wear is generally inversely proportional to the hardness "*p*" of the surface being worn away.

$$\therefore \qquad V = \frac{CLx}{P}$$

*C* is a non-dimensional constant dependent on the material in contact and their exact degree of cleanliness. If we consider two bodies sliding under an applied load "*L*" and flow pressure of the softer of the metals (*P*) then real area of contact will be given by:

$$L = P.A$$

It is assumed that the functions are same size circles of diameter *d* and hence area $\pi d^2/4$ then total number of *n* present at any instant will be give by

$$A = n.\frac{\pi d^2}{4}$$

$$\therefore \qquad n = \frac{4A}{\pi d^2} \quad \text{or} \quad \frac{4L}{\pi p d^2} \quad \text{or} \quad \frac{L}{p.\frac{\pi}{4}d^2}$$

$$\therefore \qquad N = \frac{n}{d} \quad \text{or} \quad \frac{4L}{\pi p d^3}$$

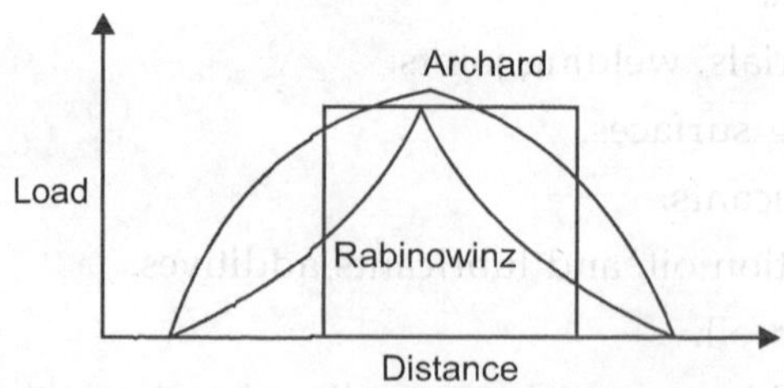


Fig. 2.12: Graph between load and distance

### Metal-Metal Adhesion

In metal-metal adhesion gold and platinum any other metal is always covered by an oxide layer film when present in unreacted form in an oxidizing atmosphere. The oxide film is to be invisible and the metal appears shiny and pure. There are basics experiments of the system, pivot and test materials with external magnetic force experiments conducted in vacuum that as the degree of surface contamination is reduced, adhesion between metallic surfaces becomes very large. In these experiments the metal was first heated to melt off the oxide film.

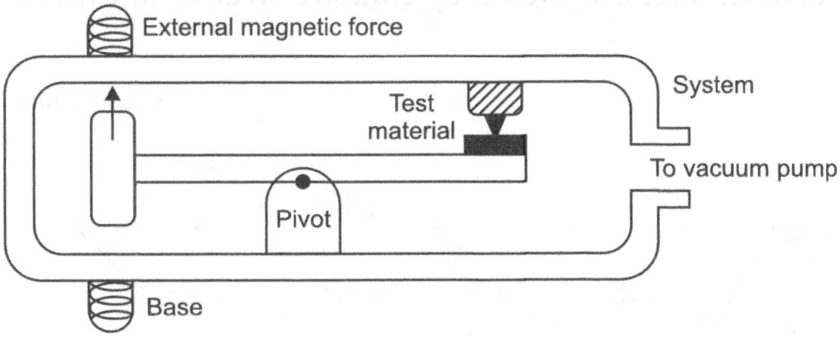


Fig. 2.13

### Metal-Polymer Adhesion

The extensive series of experiments conducted into metal-polymer contact and adhesion under high vacuum particles of the metals and polymers can also show a high degree of adhesion. Adhesion observed between a tungsten surface and polymers which are occurred on Adhesion is strong enough to cause polymer to transfer to the metallic surface when the two materials are separated.

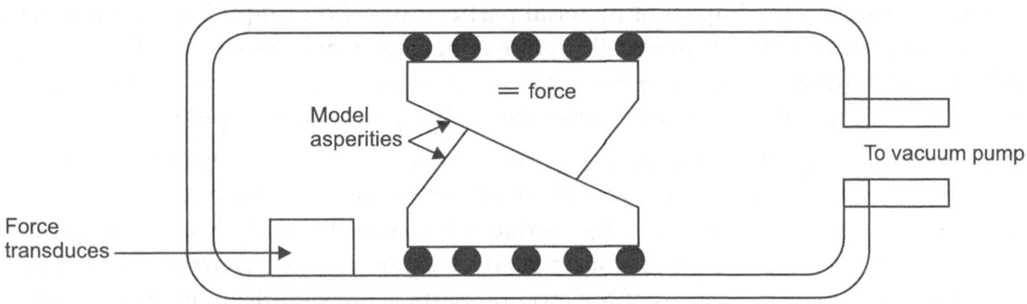


Fig. 2.14

It is known that weak Bond to moderate level of adhesion as a result of van der Waals forces acting between almost all contacting materials. Attractive forces have been found between quartz surfaces and mica surfaces. Rubber was also found to adhere to glass and polymer. In all of these cases van der Waals forces were clearly the largest component contributing to adhesion.

$$k = E\,\sigma^{3/2}\,(r^{1/2}\,\Delta\rho)$$

where

$k$ = coefficient of reduction in adhesion

$E$ = Young's Modulus ($Pa$)

$\sigma$ = standard deviation of the asperity height

$r$ = average radius of curvature of asperities.

### Control of the Adhesive Wear

Oxidation of metal surfaces can lower adhesion to acceptable levels of ceramic metals, when exposed to air, form very rapidly an oxide film of about 5 μm in thickness. A 5 μm film is so thin that it is transparent and the metal remains shiny, different changes the surface properties of the metal. When the controlling the Adhesive wear the particles. High friction with the possibility of seizure and the growth of transfer particles can result. High transfer particles can jam the sliding contact, *e.g.*, if it is annular. Problem caused by adhesive wear is an extremely high wear rate and severe surface.

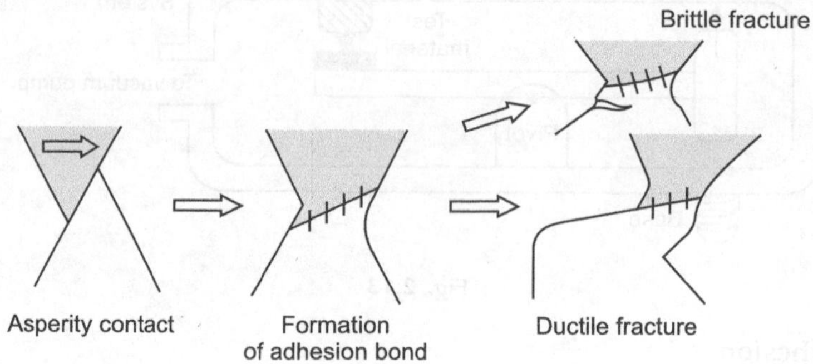


**Fig. 2.15**

### Erosive Wear

Erosive wear is caused by the impact of material particles of solid or liquid against the surface of an object. The impacting particles gradually remove material from the surface through repeated deformations and cutting actions. Erosive wear occurrs on the surfaces of bodies, the erosive wear associated with the movement of slurries through piping and pumping equipment.

Erosive wear is caused by the impact of particles of solid or liquid against the surface of an object. Erosive wear occurrs in a wide variety of machinery and the particles which are caused by the Roughly and high movement of the surfaces between them. With other forms of wear, mechanical strength wear resistance and a detailed study of material characteristics is required for wear minimization. The properties of the eroding particle are significant by increasingly being recognized as a relevant parameter in the control of wear. Erosive wear is occurred when two surface slides over each other, then the some part of the both or one surface erosive is called the erosive wear of that surfaces.

Erosive wear is caused by impingement of particles (solid, liquid or gaseous), which remove material fragments from parts or the surface due to momentum effect.

Erosive wear of **Engine part or bearings** may be caused by cavitation in the **lubrication oil**. The cavitation may form when the oil exits from the **convergent gap** between the bearing and journal surfaces. The oil pressure rapidly drop providing conditions for voids formation (the pressure is lower than the oil vapor pressure). The bubbles strikes and then collapse producing a shock wave, which removes particles of the bearing material from the bearing.

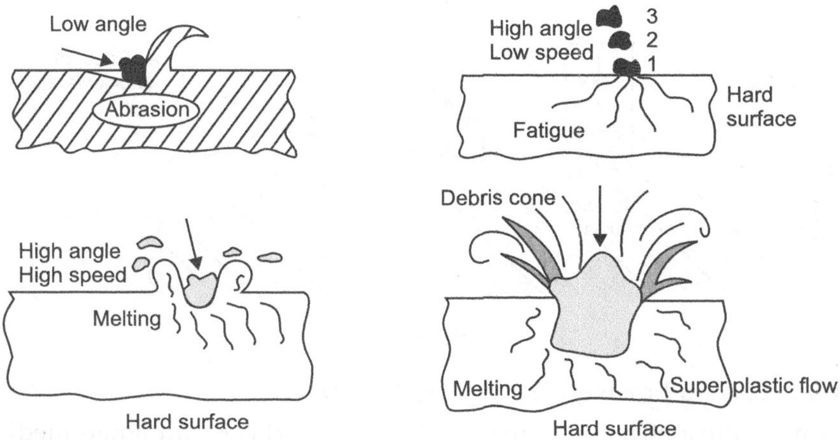


Fig. 2.16

### Mechanism of Erosive Wear

Erosive wear involves several wear mechanisms controlled by the particle material, the angle of impingement, the impact velocity, and the particle size. If the particle of material is hard and solid then it is possible that a process similar to abrasive wear will occur. Mechanism of the erosive wear when the particle size decomposes the structure of atom, bonds is empirical by nature and relates more to practical consideration to understanding of wear.

### Effect of Impingement Angle and Impact Speed on Erosive Wear Rate

Impingement angles can range from 0° to 90° at zero impingement angle there is negligible wear because the eroding particles which occurrs although even at relatively small impingement angles of about 20°, Erosive wear may occur if the particles are hard and the surface is soft based materials and structure bonds. If the surface is brittle then severe wear by fragmentation of the surface may occur reaching its maximum rate at impact angles close to 90°. When the impingement angle is 0° to 90° at erosive wear.

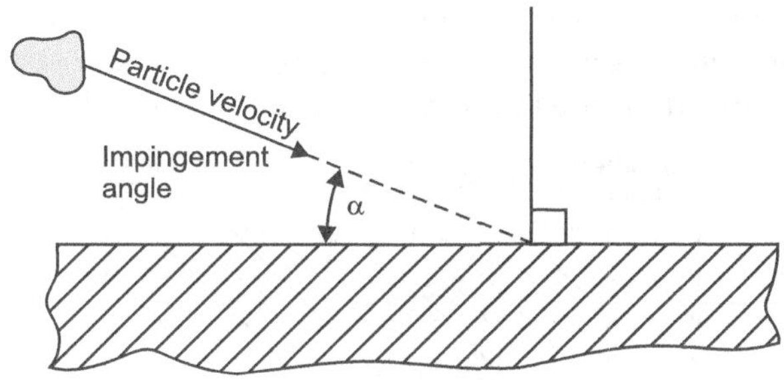


Fig. 2.17

In Fig. 2.17 shows the impingement angle between the surface when the erosive wear occurs at angle 0° to 90° angles.

### Effect of Temperature on Erosive Wear

The rate and mechanism of erosive wear are influenced by temperature. The effect of temperature is to soften the eroded material and increase wear rate. The temperature propagating the shock wave effects on the structure of the mechanism and the movement of the particles.

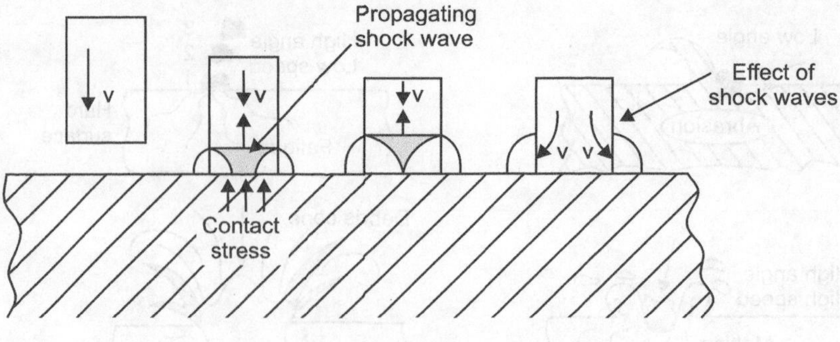


Fig. 2.18

### Effect of Erosion Media on Erosive Wear (Slurry Effect)

Erosive wear form the great effect of a mixture of erosive particles and liquid medium is known as a slurry. The characteristics of the medium have a surprisingly strong effect on the final wear rate. Controlling factors relate to the bulk properties of the medium, *i.e.*, viscosity, density and turbulence and to its microscopic properties such as corrosively and lubrication capacity. When the flow of media is transferred to the uneroded section when the particle trajectories unaffected by flow of media.

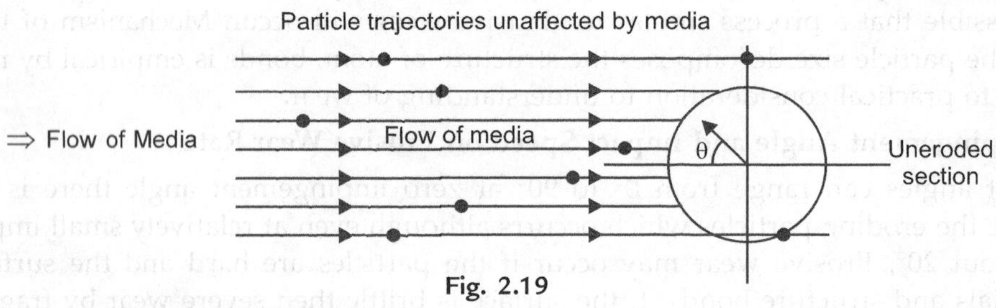


Fig. 2.19

### Erosive Wear Resistance of Materials

Resistance materials of the erosive wear which occurs in a similar manner to abrasive wear, it is found that improvements in mechanical properties always coincide with superior erosive wear resistance. When the improvements in the mechanical properties are straining of materials then the heat of deformation occurred on the surfaces then the $H_2O$ absorbed on the trapped surfaces.

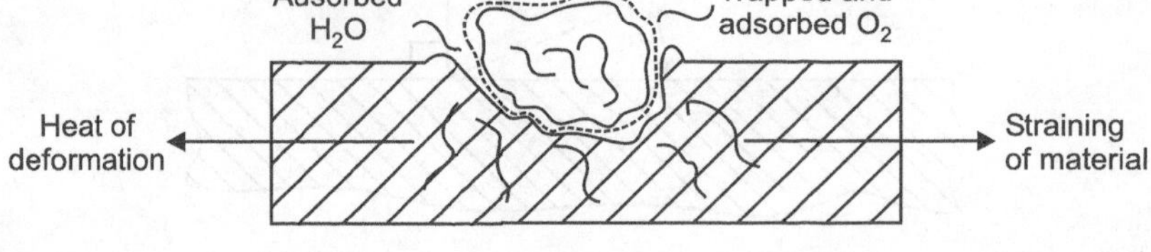


Fig. 2.20

### Cavitation Wear

Cavitation wear is known to damage equipment or surfaces on the liquid surfaces. Wear progresses by the formation of a series of holes or pits in the surface exposed to cavitation. When liquids flow parallel with a flat plate of laminar flow or turbulent flow. If liquid is to flow past a cylinder and perpendicular to the axis of the cylinder just beyond the widest dimension of the cylinder, the momentum of the liquid produces a lower pressure at the solid/liquid interface between the bond of the system. In cavitation if the radius of curvature of the cylinder is small and/or if the velocity of the liquid is high, the pressure at the solid/liquid interface may be less than the vapour pressure of the liquid. Bubbles, or cavities, of vapour will collapse very quickly. The entire machine component can be destroyed by this process. Cavitation is cyclic formation processes and collapse of Bubbles on surface.

Liquid is the medium that causes cavitation wear. Cavitation wear does not require a second surface; it requires only that high relative motion exists between the surface and the fluid. Which motion reduces the pressure in the fluid. When the liquid reaches its boiling point and ebullition occurrs, vapour bubbles form, which produces cavitation, vapour cavity lasts a short time because almost any increase in pressure causes the vapour in the bubble to condense instantaneously and the bubble to collapse and produce a shock wave. This shock wave then impinges on adjacent metal surfaces and destroys the material bonds according to movement of liquids particles.

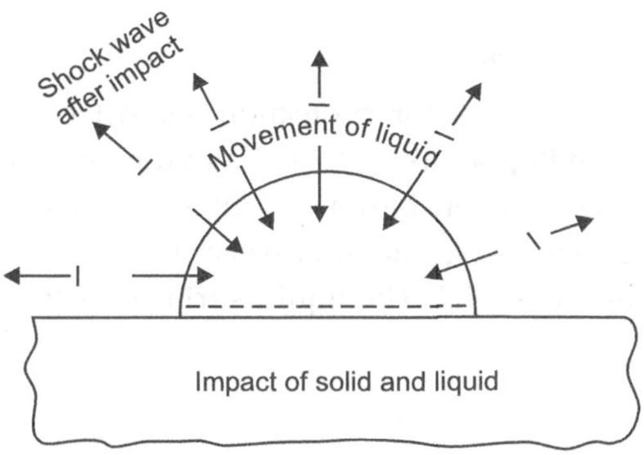


Fig. 2.21

### Mechanism of Cavitation Wear

The characteristic of cavitation is the cyclic formation and collapse of bubbles on a solid surface in contact with a fluid. Bubble formation is caused by the release of dissolved gas from the liquid where it sustains effect of negative pressure. Negative pressure is likely to occur when flow of liquid enters a diverging geometry, *i.e.,* emerging from a small diameter pipe to a large diameter pipe. The effect of the sharp sided object moving in liquids, *e.g.,* ship propeller, is particularly prone to cavitation. The ideal method of preventing cavitation is to avoid negative pressure close to surface, but in practice this is usually impossible. When a bubble collapses on a surface the liquid adjacent to the bubble is at first accelerated and then sharply decelerated as it collides with the surface. The collision between liquid and solid generates large stresses which can damage the solid and the surfaces of the object, cavitation wear occurs on the surface of the liquid where the bubbles strikes on the surfaces due to the gasses occurs in the bubbles when the gases released from the liquid if sustains the near zero pressure.

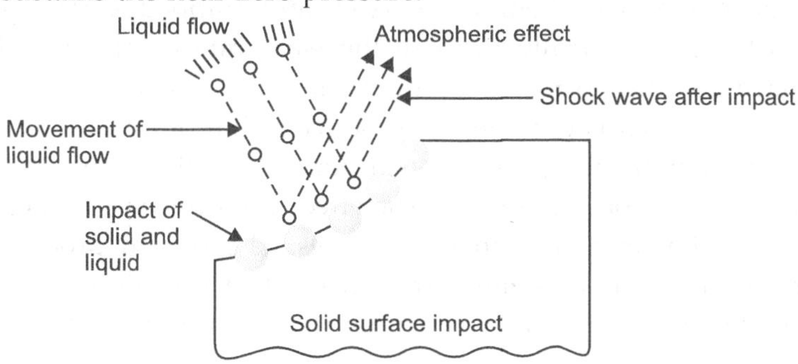

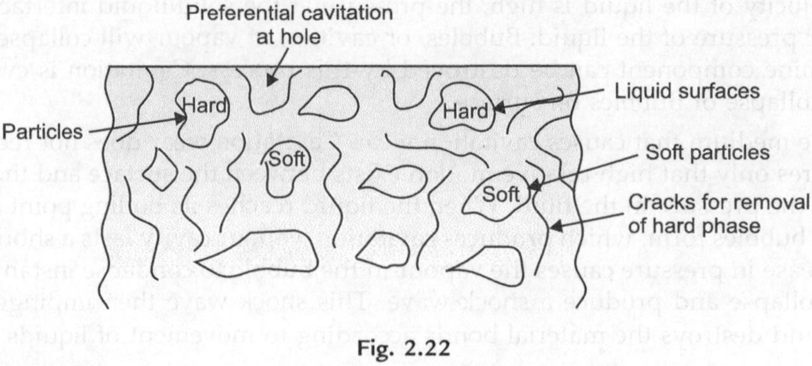


Fig. 2.22

## Cavitation Wear Resistance of Materials

Wear Resistances the choice of material for protection against cavitation wear is the physical contamination scale where the cavitation takes place. Cavitation can occur in components ranging from propellers to dam spillways, concrete based materials are often used, for the cavitation wear resistance materials which occurs on surfaces. A basic feature of cavitation is its preferential attack on the weakest phase of a material. The liquid surfaces of the soft and the Hard particles collides on each others.

## Corrosive Wear

Wear may be accelerated by **corrosion (oxidation)** of the rubbing surfaces when mechanical action takes place. Increased temperature and removal of the protecting oxide films from the surface during the friction promote the oxidation process. Friction provides continuous removal of the oxide film followed by continuous formation of new oxide film. Corrosive wear occurs when the surfaces is formed corrosive.

$$Fe + H_2SO_4\,(aq) \rightarrow FeSO_4 + H_2$$

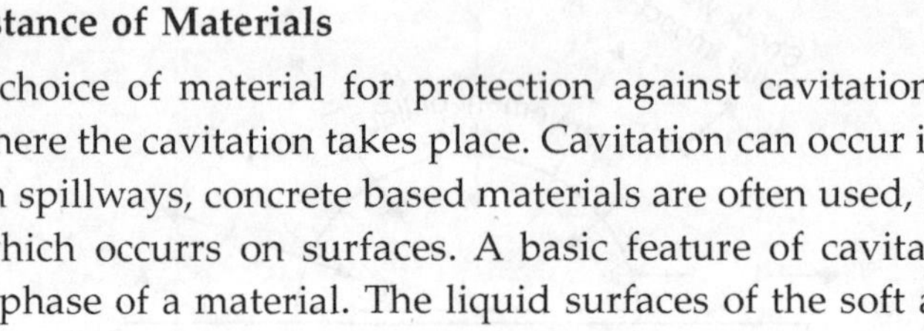


Fig. 2.23

Hard oxide particles removed from the surface and trapped surface particles between the sliding/rolling surfaces additionally increase the wear rate by **three body abrasive wear** mechanism. Thick layer of the corrosive products and the corrosive forming layer on the surfaces of metals. If a material is corroded to produce a film on its surface. The first process is the formation of durable lubricating film. If such film prevail then the worn contacts are well lubricated and corrosive wear does not occurs. The effect of corrosion product films are durable so that this category of film formation on the surfaces. The second process is related to the formation of a sacrificial or short life-time corrosion product film under sliding contacts. The common form of corrosive wear since most corrosion film consist of brittle oxides.

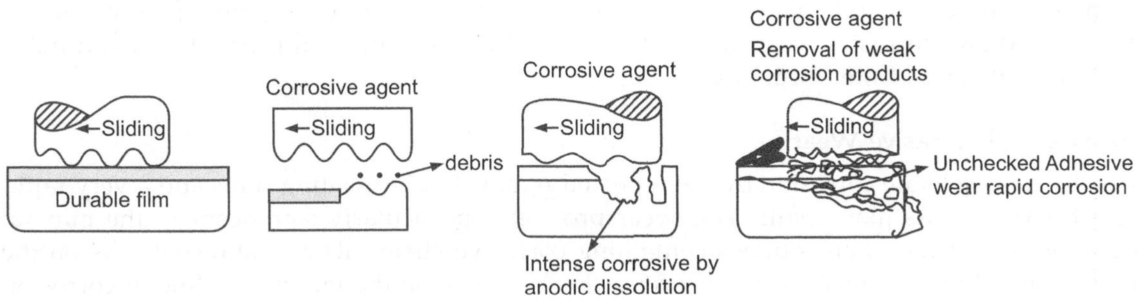


**Fig. 2.24**

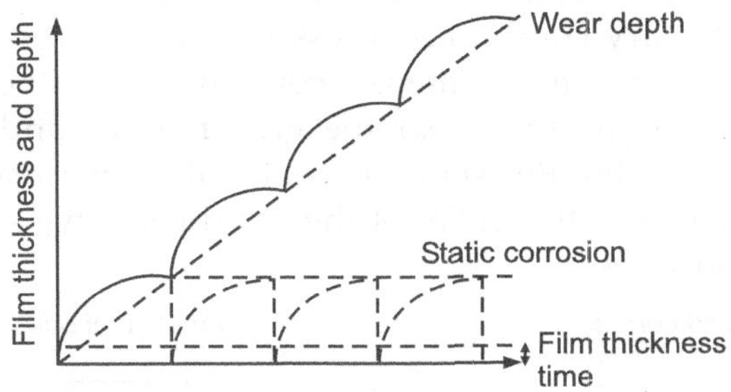


**Fig. 2.25**

The surface chemical reactions which are beneficial in preventing adhesive wear lead to a considerable loss of the underlying material. If a material is corroded to produce a film on its surface.

- a durable lubricating film which inhibits both corrosion and wear may be formed on surfaces.
- a weak film which has a short life-time under sliding contact may be produced at high rate of wear may occur due to regular formation and destruction of the films. The friction coefficient may or may not be low in this instance.
- the protective surface films may be worn between the remaining films underlying substrate may result in rapid corrosion.
- the corrosive and wear processes may act independently to cause a material loss on the surfaces and missallignment of the surface.

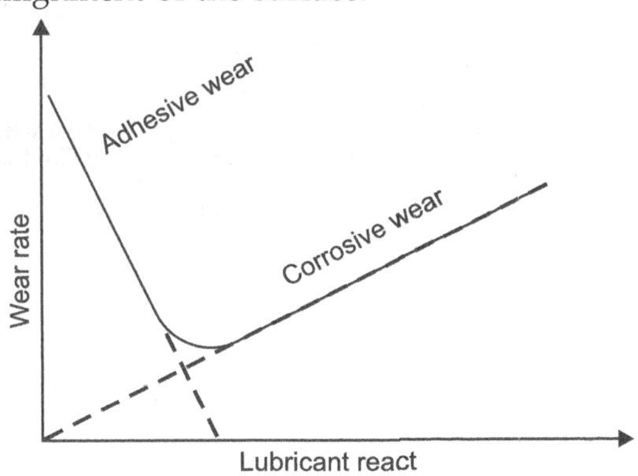


**Fig. 2.26**

Figure 2.26 show the relation between the wear rate and the lubricant reaction, where the graph plots the deviation between the adhesive wear and corrosive wear.

The wear process is dominated by the formation of durable lubricating film. If worn contacts are lubricated and corrosive wear does not occurrs. Very few corrosion product film is durable so that this category of film formation.

### Corrosive and Abrasive Wear

Abrasion can accelerate corrosion by the repeated removal of lubricating films and a very rapid form of material loss may result. This wear process is particularly significant in the mineral processing industries where slurries containing corrosive chemicals and abrasive grits on the surfaces when the movement or the process which occurrs on the fact that although corrosion was occurring the surface remained untarnished due to the prompt removal of corrosion products was provided on the surface or between the bodies. This mechanism of wear rate of mechanical abrasion under dry conditions is less than the corrosion rate without abrasive wear. When mechanical abrasion is more intense, corrosive effects become insignificant when corrosion is slow compared to abrasive wear the grits remove underlying metal with little interference from the corrosion film. Resistance of materials under corrosion-abrasion depends on their resistance to corrosion and the initial of the rapid corrosive surfaces. The formation of the passivation film on the surfaces.

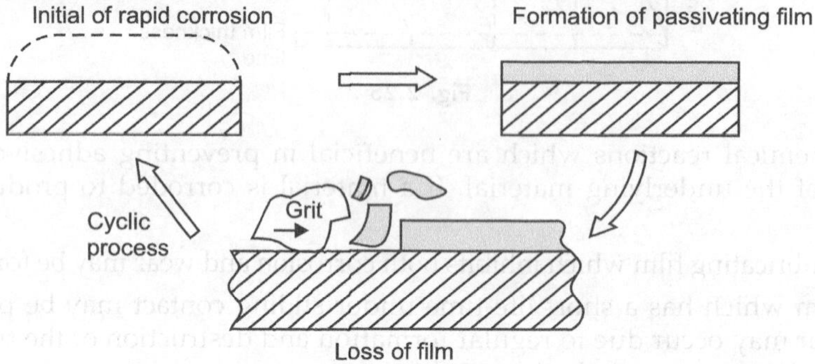


Fig. 2.27

According to Fig. 2.27 the initial of rapid corrosion formates the formation of passive film, the most common form of corrosive films consist of brittle oxides or other ionic compounds. The oxides of iron are extremely brittle at all but very high temperatures.

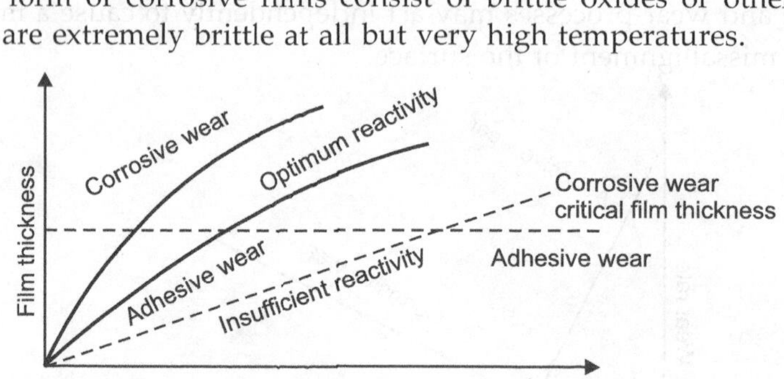


Fig. 2.28

Figure 2.28 show the relation between the film thickness and time so that when the time is increases then the thickness of film also increases but after some times it comes to down on the critical film thickness to the insufficient reactivity.

Wear in highly corrosive media while effectively limited to extremely corrosive media where the corrosion products weak and are probably soluble in the liquid media. When wear and corrosion, if occurring in the same system, can proceed entirely independently since the heat and mechanical agitation of a sliding contact would almost inevitable accelerate corrosion and crack formation and propagation on the surfaces.

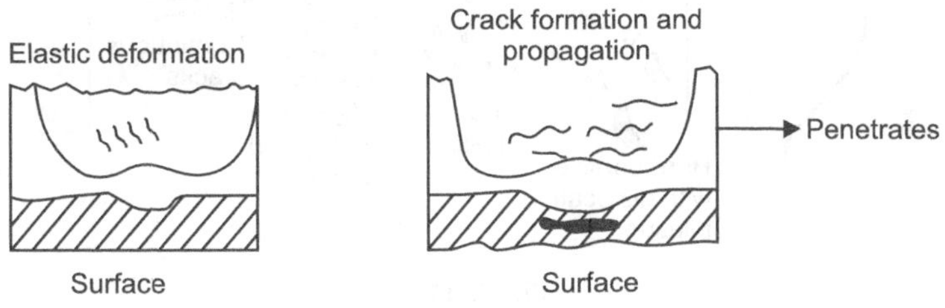


Fig. 2.29

## Fatigue Wear

Fatigue wear of a material is caused by a cyclic loading during friction. Fatigue wear occurs if the applied load is higher than the **fatigue strength** of the material. Fatigue cracks start at the material surface and spread to the subsurface regions. The cracks may connect in separation delamination of the material pieces. **Fretting wear** caused by cycle sliding of two surfaces across each other with a small amplitude (oscillating). The friction force produces alternating compression-tension stresses, which result in surface fatigue. This wear is caused by deformations sustained by the asperities and surface layers when the asperities of opposing surfaces make. Contacts between asperities accompanied by very high local stresses are repeated a large number of times in the mechanical action of wear particles are generated by fatigue propagated cracks. Wear under these conditions is determined by the mechanics of crack initiation, crack growth and fracture as shown in Fig. 2.30 the surface is being cracked when the fatigue occurred.

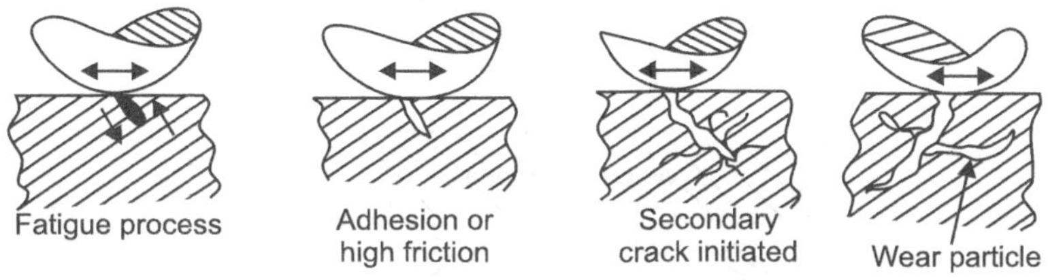


Fig. 2.30

## Surface Crack Initiated Fatigue Wear

Cracks and blocks have been observed on micrographs of worn surfaces. The mechanism of surface crack initiated fatigue wear on the surfaces of bodies, crack originates at the surface and propagates downward along weak planes such as slip planes or dislocation cell boundaries. Surface fatigue is a process by which the surface of a material is weakened between and the edges of surfaces. Fatigue wear is produced when the wear particles are detached by cyclic crack growth of micro cracks on the surface. These micro cracks are either superficial cracks or subsurface cracks generated on the surface of the object.

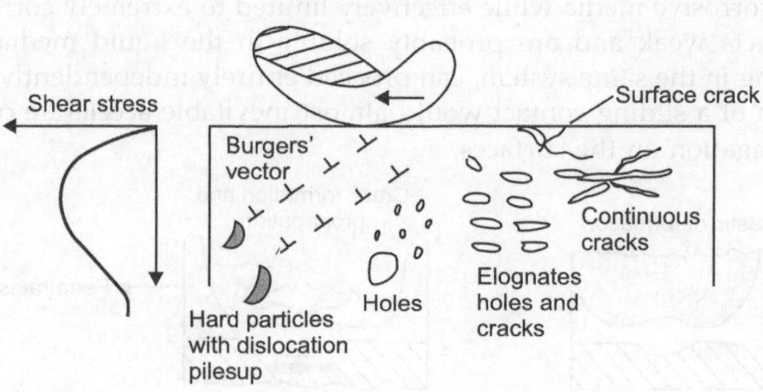


Fig. 2.31

**Fatigue Wear During Rolling**

During rolling the local contact stresses are very high, concentrated, repetitive, and wear mechanisms are determined mostly by material and the motion of the object, lubrication wear of metals or other oxidizable solids proceeds by the cyclic fragmentation of oxide films and their subsequent reformation. In other words, a form of oxidative wear occurs. In Fig. 2.32 shows that the rolling mechanism on the wearing surfaces effect the material removal.

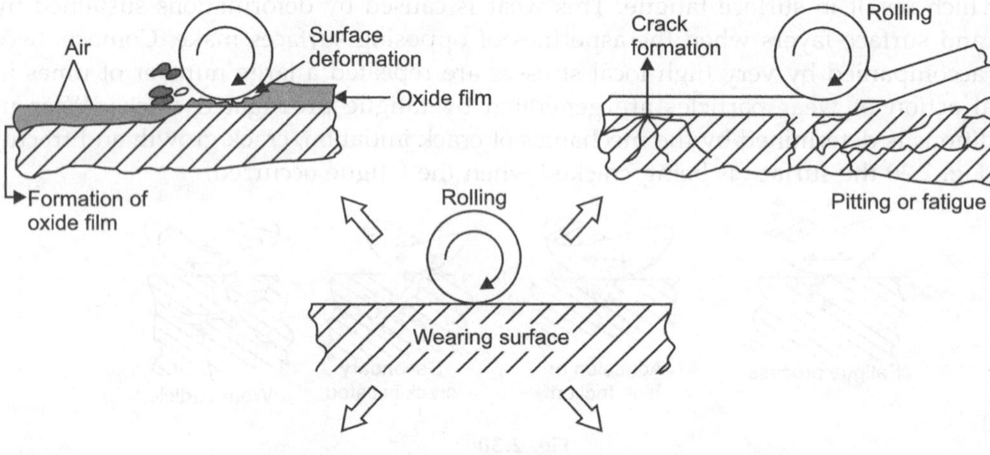


Fig. 2.32

**Fretting Wear**

Fretting wear is the repeated cyclical rubbing between two surfaces, over a period of time which will remove material from one or both surfaces in contact. It occurs typically in bearings, have their surfaces hardened to resist the problem. When cracks in either surface are created, known as fretting fatigue. It can lead to catastrophic failure of the bearing. An associated problem occurs when the small particles removed by wear are oxidized in air. The oxides are usually harder than the underlying metal, so wear accelerates as the harder particles abrade the metal surfaces further. Fretting corrosion acts in the same way, especially when water is present.

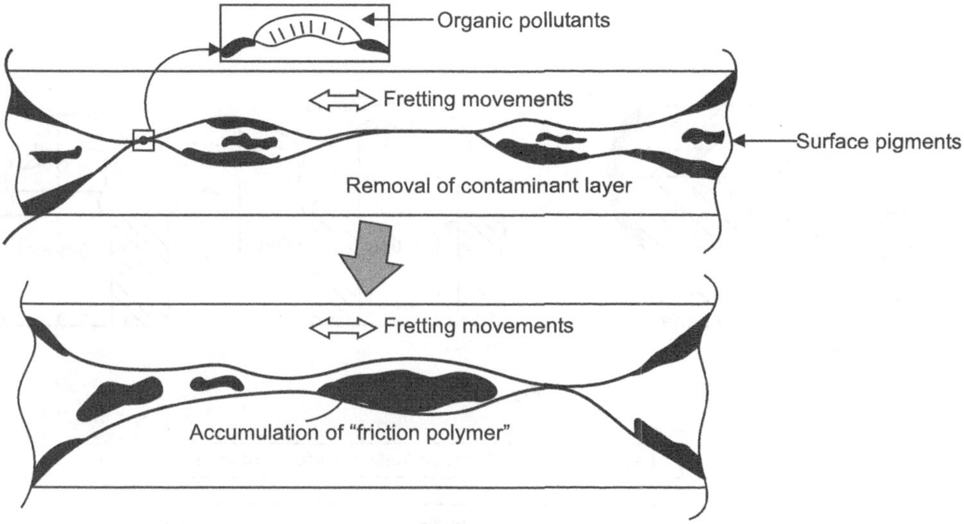


Fig. 2.33

When two solids surfaces are pressed together subjected to a tangential force of increasing magnitude, there is a certain value of tangential force at macroscopic sliding occurrs. In the surface pigments it is less widely realized that at levels of tangential force below this limiting value tangential micro movements also occur in response to the applied force when the load is applied on the surface.

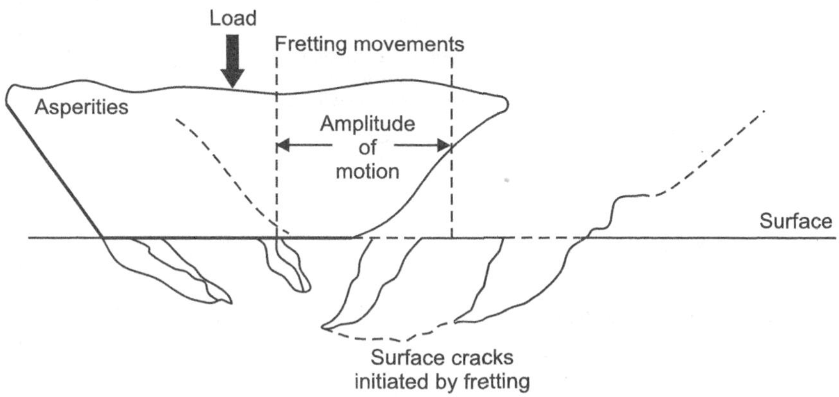


Fig. 2.34

### Environmental Effects on Fretting Wear

Due to the moisture atmospheric oxygen and water have a strong effect on the fretting process, especially in metals. Environment oxygen and water accentuate fretting wear and surface damage and that an inert atmosphere such as argon or nitrogen suppresses fretting wear from metallic contacts is high in oxide content and consists of a fine powder. Materials like iron or steel, the debris often shows a reddish-brown colour characteristic of iron oxide, for aluminium the debris is black.

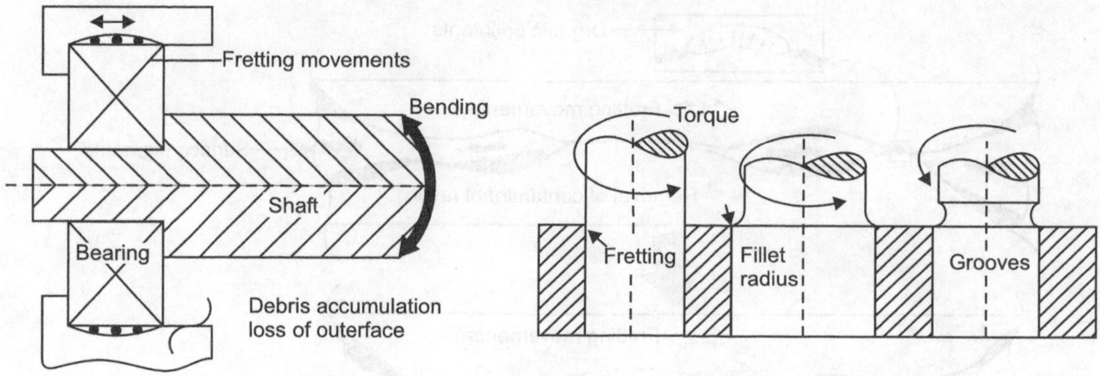


Fig. 2.35

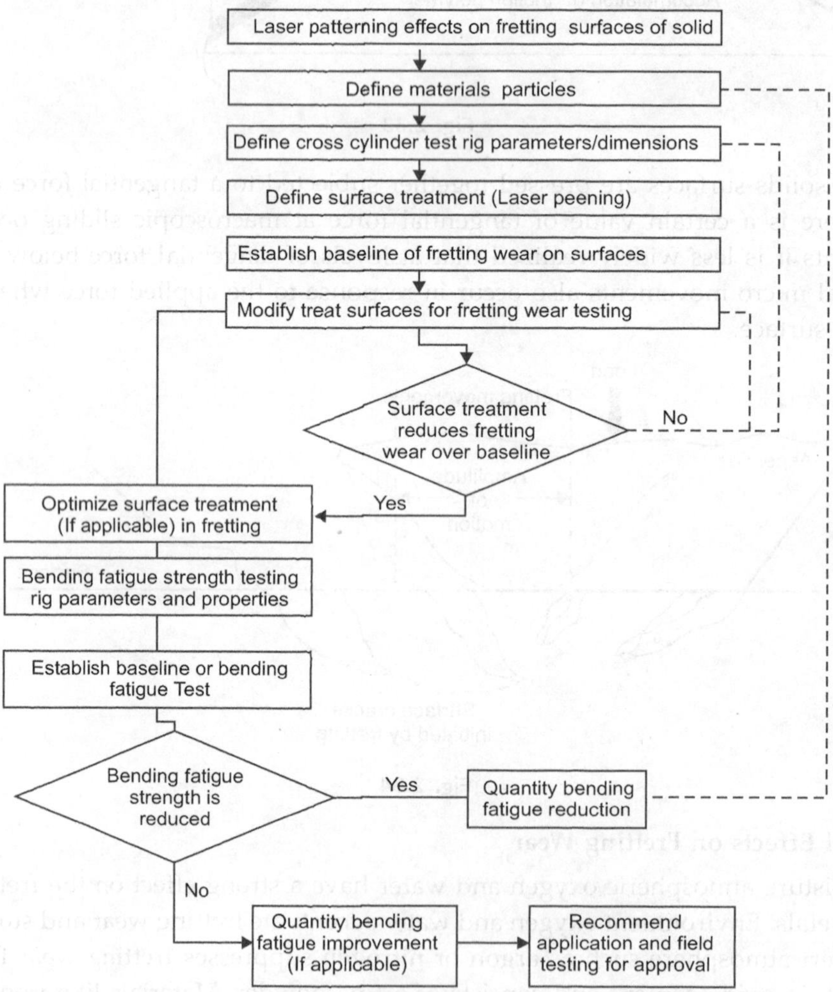


Fig. 2.36

Fretting wear usually begins at a high rate cycles of the effects cycles in steel, depending on the ductility of the oxide builds around small effect contact areas and carries much of the load. The debris from steel is usually the red, nonmagnetic, hexagonal $Fe_2O_3$ if the temperature has not exceeded 200°C. At high contact pressure the oxide may appear black blocks on surfaces and the μ of this form may be low. At higher temperature, the fretting rate decreases.

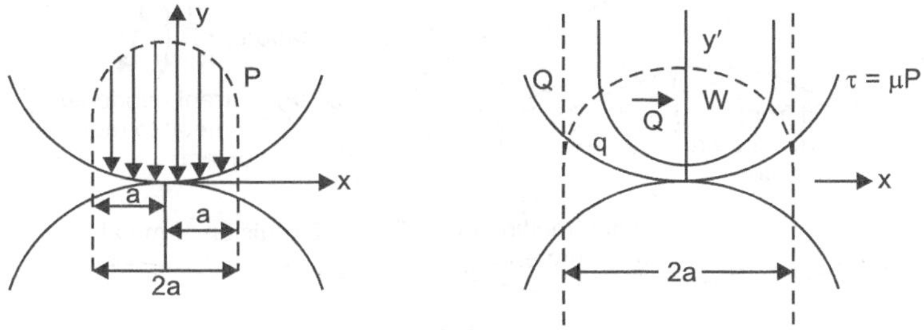


Fig. 2.37

### Effects of Temperature and Lubricants on Fretting

Temperature may affect on the lubricating film the corrosion and oxidation rates usually increase with temperature, and the mechanical properties of materials change with temperature. In metals the temperature effects on fretting in terms of surface oxidation kinetics. Fretting wear rates usually decrease with increasing temperature if a stable and adherent oxide film is formed on the surface. Decrease in fretting with temperature increase may be quite surprising since the oxidation rates of steel increase with temperature.

### Melting Wear

Melting wear since any material which melts without decomposing of material particles will show melting wear under appropriate conditions. The present concept of melting wear is based on the phenomenon that when sliding speed and load is increased, the frictional surface temperature also rises until the melting temperature is reached and layers of molten surface material begin to influence friction and wear.

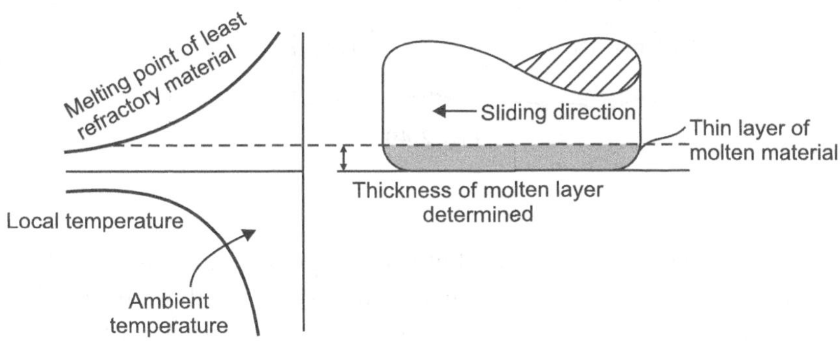


Fig. 2.38

The sliding conditions often result in the formation of hard particles white layers on the worn surface. These layers do not etch to a microstructure and simply appear as white areas under the metallurgical microscope. In some cases these layers form after rapid solidification of the molten top surface. For example, in gun barrels, where there is a combination of high stress, high temperature and a hostile environment, the white layers are formed by rapid solidification of molten material. Figure. 2.39 explains the position of body moment on the arc formation.

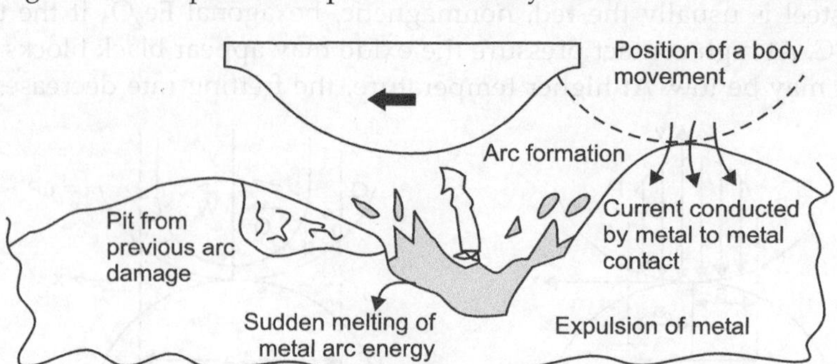


Fig. 2.39

## Diffusive Wear

In diffusive wear when there is true contact between the atoms of opposing surfaces and a high interface temperature, significant diffusion of chemical elements from one body to another body movements. Contact is the rake face of a cutting tool close to the cutting edge in high speed machining. Processes the perfect contact between the tool and the metal chip due to the extreme contact stresses and very high temperatures, occurred. The metal chip represents a continually supply of relatively pure metal whilst the tool is a high concentration mixture of some radically different elements, *e.g.*, tungsten and carbon is a tendency for some of the elements in the tool to diffuse into the chip where solubility conditions are more favorable.

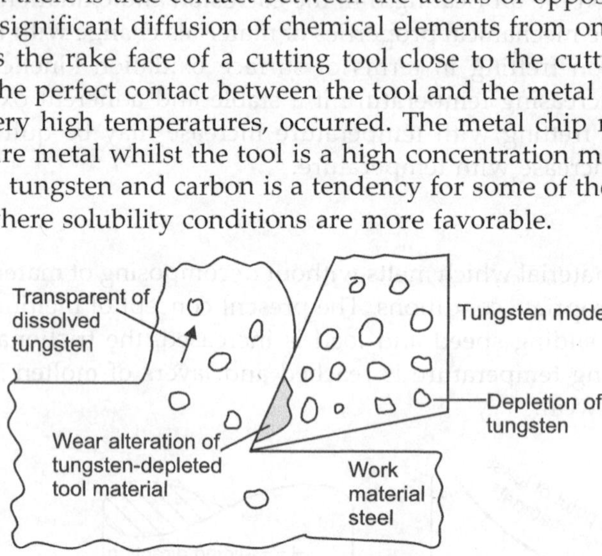


Fig. 2.40

## Impact Wear

Impact wear is caused by repetitive collision between opposing surfaces and the great effect on the surfaces of metals. This form of wear involves flat surfaces with a large radius of curvature compared to the size of the wear scan. The impact wear from erosive wear where a sharp particle indents a flat surface. In impact wear the surface is subjected to repetitive impact by a series of pulses of high contact stress combined with some energy dissipation. The mechanism of impact wear involves elastic and plastic deformation when impact energy is high and/or fatigue accompanied by wear debris release due to crack formation. If oxygen is present and the wearing material can be oxidized then a corrosive or oxidative wear mechanism can also take place. When the load or other surfaces pull load on the another surfaces then the impact occurs between the surfaces.

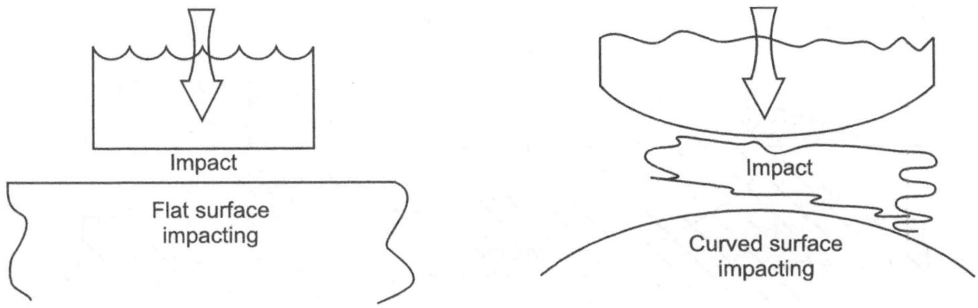


Fig. 2.41

The type of impact wear has a strong effect on the wear mechanism where metals are prone to an oxidative form of impact wear while ceramics wear by cracking the spalling. Polymers tend to wear by plastic deformation, fatigue cracking or by chemical attack from hot compressed layers of oxygen and pollutants at the moment of impact on the surfaces.

## 2.3 TRIBOLOGY OF POLYMERS

The term 'polymers', which is flat surfacing is formed by the polymerization of hydrocarbons, is used to describe an enormous range of different substances. These materials were never intended to be utilized as bearing or wear-resistant materials. A few polymers, however, do have valuable tribological properties and most research is directed towards this relatively limited number of polymers.

### Wear of Ceramic Materials

*General Features of Wear*: There are four fairly consistent differences between metals and ceramic materials in sliding contact:

1. The coefficient of friction of ceramic materials is significantly higher than that of metals. The wear of ceramic materials is based on the effect of material or surfaces vibratory effect.
2. In repeat-pass sliding with the pin-on-disk shape, the wear loss from the pin is greatest for metal combinations (unless the disk is much softer than the pin), whereas the wear loss from the disk is greatest for ceramic combinations. When the body sliding then the wear rates rises.
3. The wear rate often increases sharply at some point during an increase in sliding speed, probably due to thermal stress cycling.
4. The wear rate often increases sharply at some point during an increase in contact pressure. Ceramic materials are different from metals and polymers in two very important respects that influence wear and surface damage.

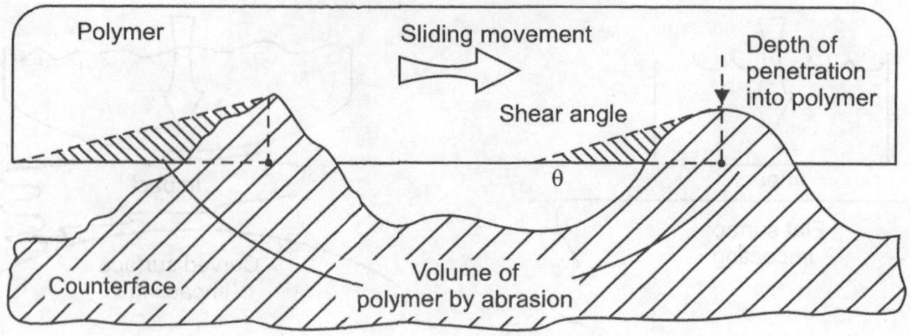


Fig. 2.42

| Polymer | Tribological characteristics |
|---|---|
| PTFE Polytetrafluroethylene | Low friction but high wear rate, usually blended with other polymer bonds |
| Nylons | Moderate coefficient of friction and low wear rate characteristics |
| Polyacetals | Performance similar to nylon, PVC |
| Polyimides | High performance polymers, for high contract stress |
| Epoxies and phendics | Used as binders in composite material and particles |

The mechanism of wear characteristics form of the wear kinetics is believed from the periodic release of molten polymer from the wearing contact. When the temperature is too low to sustain continuous melting, melting proceeds by a cycle of gradual formation of molten polymer. The molten material when the entire wear surface is covered by molten polymer.

**Friction and Wear in the Rubbery State**

Rubbers and rubbery materials can provide a unique combination of low wear and high friction coefficients and are widely used for pneumatic tyres and pipe-lining. Rubbers or polymers have a molecular structure which allows extremely large strains to be imposed before fracture occurrs. The rubber or elastomeric consists of long linear molecules which obtained to form an amorphous solid. When the material is strained, the molecules untangle and align in the direction of strain.

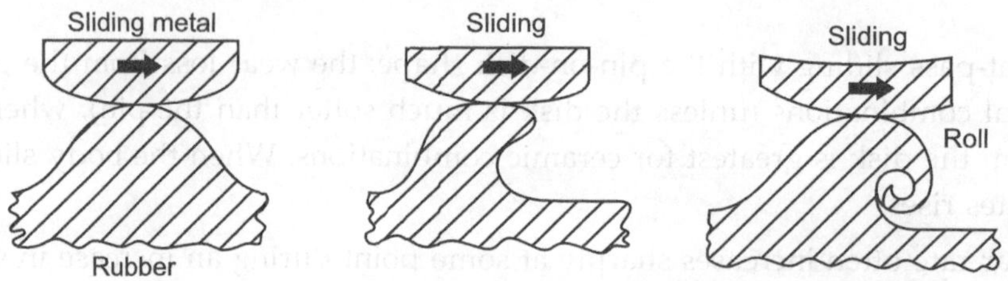


Fig. 2.43

## 2.4 TRIBOLOGY OF POLYMER COMPOSITES

Polymers are very rarely used as bearing materials even a nominally pure polymer, *e.g.*, nylon, contains plasticizers and colouring agents. The wear of polymers is strongly influenced plasticizers cause a reduction in friction for polyethylene by diffusing to the polymer surface to form a lubricating layer of wide variety of composite materials available, but the basic types of polymer composites are:

- bulk polymer containing a lubricating polymer, *e.g.*, nylon + PTFE,
- polymer containing a metal or an inorganic powder, *e.g.*, PTFE + lead powder,
- metal wire or inorganic fibre reinforced polymer, *e.g.*, glass reinforced PTFE,
- reinforced polymer containing a second lubricating polymer, *e.g.*, glass reinforced nylon + PTFE.

**Polymer Blends**

When add a polymer, usually PTFE, to another polymer in order to reduce the coefficient of sliding friction while maintaining a low wear rate. Example of these materials are nylon or polyacetal with added PTFE. The tribological characteristics of these polymer composites appear to depend on the PTFE fraction depositing a thin transfer film on the counter face.

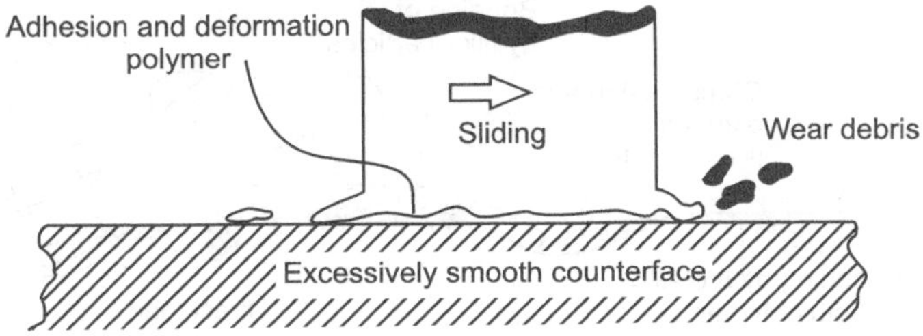


**Fig. 2.44**

**Fibre Reinforced Polymers**

Fibre reinforced polymers are a very important type of composite and a wide range of materials belong to this category. There are two basic forms of fibre-reinforced polymer:

- polymers reinforced by randomly oriented chopped fibres, PVC, polyvinyal.
- polymers reinforced by unidirectional or woven fibres on fibres.

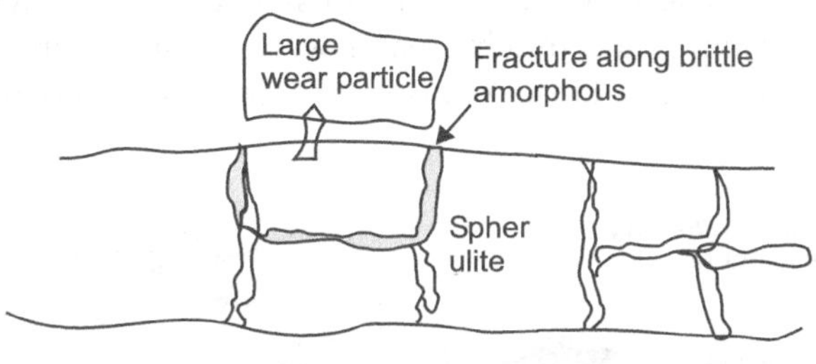


**Fig. 2.45**

Reinforcements are effective in reducing wear provided that there is a strong adhesion between the fibres and the matrix found that the limits of sliding speed and contact stress can be raised and the wear rate lowered by the incorporation of chopped fibres.

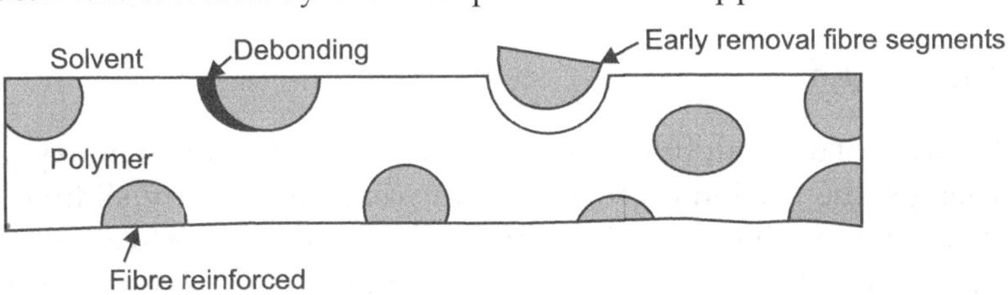


**Fig. 2.46**

**Powder Composites**

A wide range of powdered metals and inorganic solids added to polymers. The powder reinforcements in polymers include: copper, lead, zinc, non-ferrous alloys such as bronze and solid lubricants, such as graphite and molybdenum disulphide. Powdered glass has also been used as a reinforcement material which are generally the addition of solid lubricants to polymers results in some improvements in their wear and frictional characteristics. The degree of these improvements varies greatly depending on the type of polymer and lubricant used between particles. The polymers of moderate lubricity such as nylon and polyimide exhibit the greatest improvements with the addition of solid lubricants, while the improvements to PTFE are very limited.

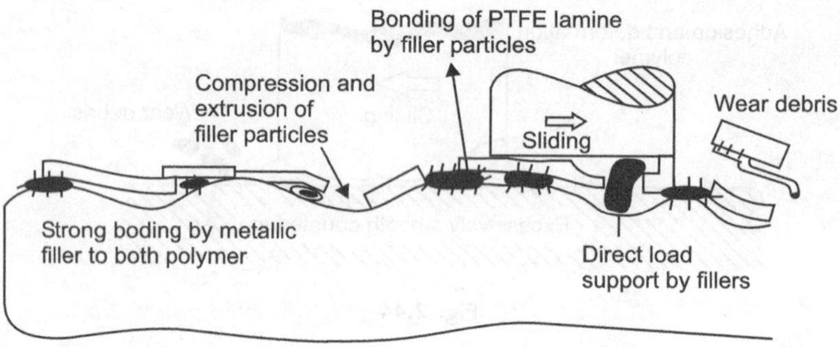


Fig. 2.47

**Wear and Friction of Ceramics**

Ceramics as materials that include a wide range of hard refractory inorganic compounds, which are formed by heating the base material in powder form to a high temperature where sintering or solid state reaction occurs. The result of this process is a material which possesses superior hardness, good chemical resistance and much greater wear resistance than most metals. Ceramics at ambient temperature revealed a close relationship between maximum tensile stresses in wearing contact.

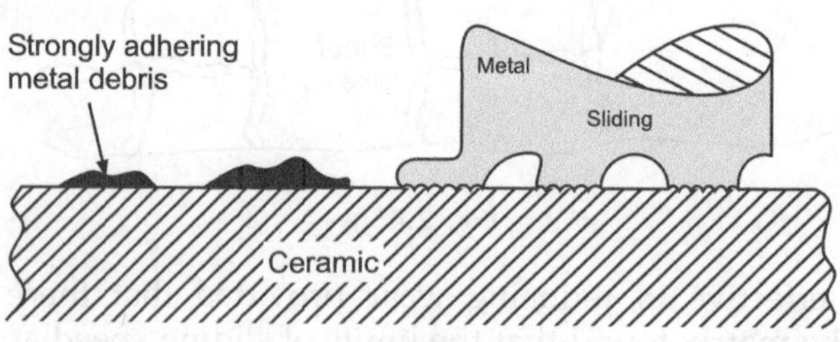


Fig. 2.48

## 2.5 OXIDATIVE WEAR

Oxidative wear is the wear of dry unlubricated metals in the presence of air or oxygen. Atmospheric oxygen radically changes the friction coefficients and wear rates of dry sliding metals involved in the process. Oxidative wear changes in the chemical composition of wear debris generated in dry sliding of steels under different levels of load and sliding speed were observed.

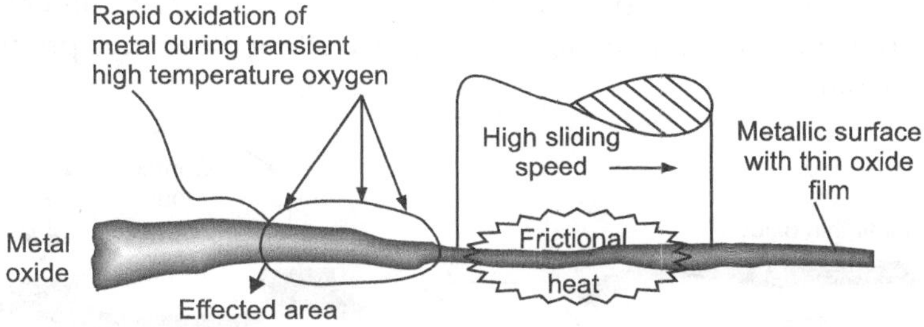


Fig. 2.49

When the load and sliding speed were high enough to increase the frictional contact temperature to several hundred degrees Celsius, the wear debris changed from metallic iron to iron oxides. When thick oxide films were formed on the worn surfaces then 'mild wear' Oxidative or mild wear shows a moderate and stable coefficient of friction of about **0.3–0.6** compared to much larger fluctuating values for severe wear. The characteristic features of oxidative wear are smooth wear surfaces and small oxidized wear debris.

At low temperatures, the oxide films are extremely beneficial since they form rapidly and effectively suppress adhesive wear. If a system operates under mild oxidational conditions wear is greatly reduced. The oxide layers formed by the strain-hardened substrate layers which are generated due to plastic deformation. At high temperatures, oxidation resembles corrosion in its high rate of reaction and can become a direct cause of increased wear. This rapid oxidation at high temperatures forms the base of oxidative wear.

In Fig. 2.50 shows that the graph between oxide film thickness and time. When the time increases the high temperature increases and low temperature oxidation decreases.

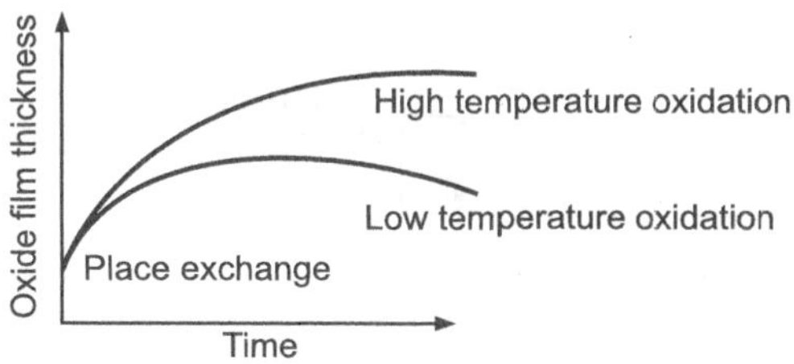


Fig. 2.50

### Means of Controlling Corrosive and Oxidative Wear

The addition of corrosion inhibitors to the lubricating oil or process fluid can be an effective means of controlling corrosive wear. However, in lubricated systems there is a significant risk of interfering lubrication additives so that ad hoc addition of corrosion inhibitors to lubricating oils in unwise. The corrosion inhibitors can function by the adsorption or the formation of a passivating layer on the operating surface.

This layer acts as a barrier which prevents oxygen and water from reaching the surface. The corrosion inhibitor of the materials adsorbed layers of lubricants and promote adhesive wear. E.P. lubrication, which is a form of controlled corrosion, may also be disrupted by corrosion inhibitors. Most oils are used to lubricate a number of contacts, which are operate under different loads and sliding speeds, Consequently if, for example, a corrosion inhibitor is added to suppress corrosive wear in one contact, much more severe adhesive wear may be initiated in an adjacent contact.

Where there is no danger of disrupting lubrication, *e.g.*, in a contact immersed in the process fluid, then the addition of corrosion inhibitors to the wetting fluid can produce a significant reduction in the wear rate.

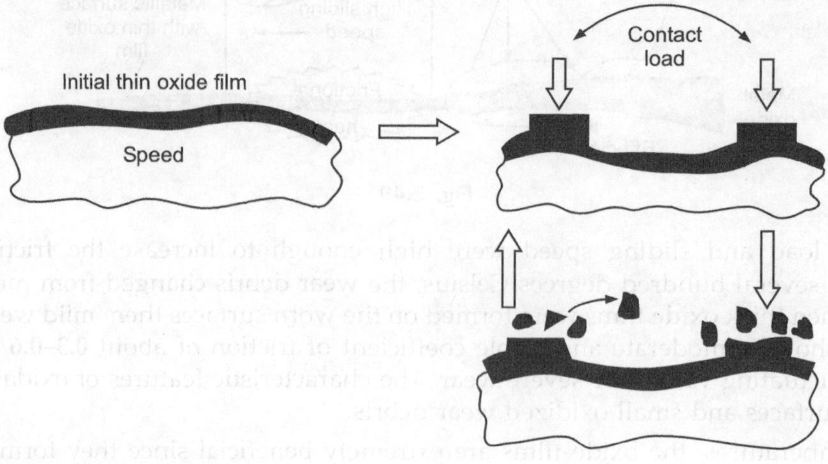


**Fig. 2.51**

## 2.6 METHODS OF WEAR MEASUREMENT

**1. Optical Method:** The ability to measure wear or erosion in real time with very high sensitivity can dramatically accelerates produce development programmes are trouble shooting. Wear/Erosion can be followed either by monitoring the removal of marker or by measuring the accumulation of labelled debris.

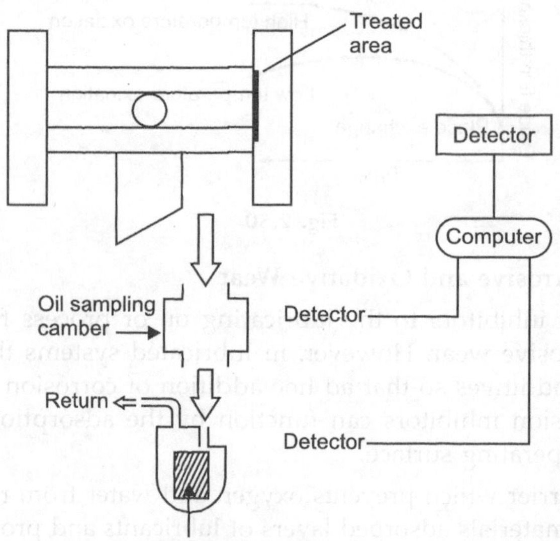


**Fig. 2.52**

**2. Stylus Profiler:** Stylus profiler has resolution ranging from 25–50 nm while optical profiler resolution ranging from 0.5–2 nm with the help of fully automated profiler, three dimensional worm surface profiles can be obtained. It is measuring instrument used to measure a surface's profile, in order to quantify its roughness.

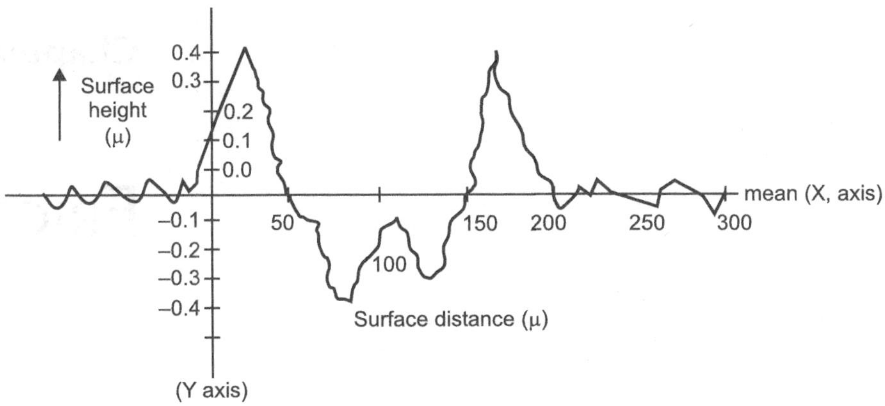


Fig. 2.53

**3. Radio Active Decay:** The activity of the auto segments produced by means of scientillation detector which particularly sensitive to γ-rays. The isotopes which in new-peak reactor emits both β-rays radiation is stopped by an interposed at plates.

$$R = R^* \frac{D - N}{D^* - N^*}$$

**4. Weight Loss Techniques:** Weight loss technique is not effective in the case of thin wear resistant coating, where the amount of wear is small while is preferable to measure large amount of wear.

**5. Ferrography/Electron Microproble Analysis :** With an electron microscope analysis a very small localised area can be investigated.

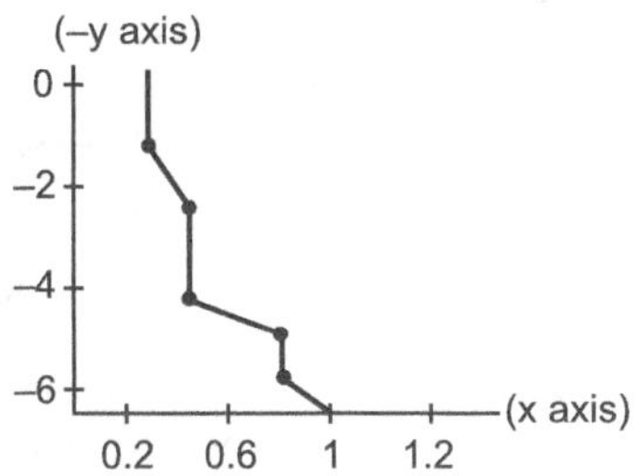


Fig. 2.54

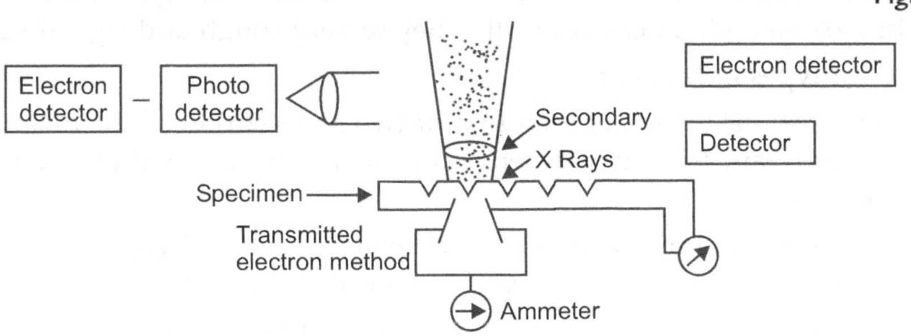


Fig. 2.55

Chapter 3

# FRICTION

## 3.1 INTRODUCTION

**Friction** is the force resisting the relative motion of solid surfaces, fluid layers of particles or material elements sliding against each other when the surface are same or different materials. Friction is the "evil" of all motion. No matter which direction something moves in, friction pulls it the other way. Move something left, friction pulls right. Move something up, friction pulls down. It all depends on the movements of the objects. It appears as if nature has given us friction to stop us from moving anything from one side to another.

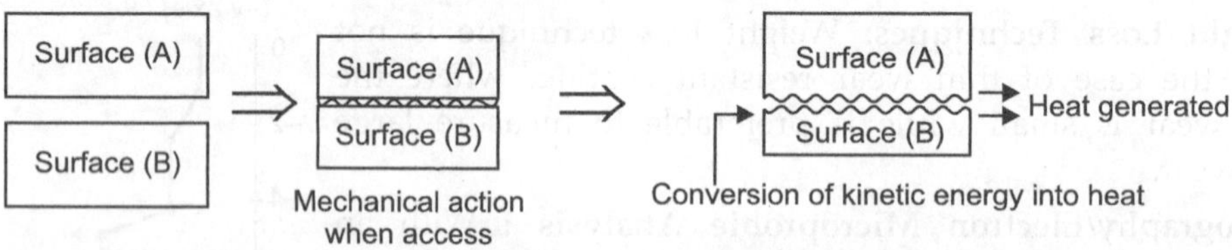


**Fig. 3.1(A):** Shows when the relative motion occurs between the solid surfaces, how heat is generated when friction between the sliding surface.

Friction is actually a force that appears whenever two things rub against each other. Although two objects might look smooth, microscopically, they're very rough and jagged surfaces.

There are several types of friction:

- *Dry friction* resists relative lateral motion of two solid surfaces in contact. Dry friction is subdivided into static friction between non-moving surfaces, and kinetic friction between moving surfaces.
- *Fluid friction* describes the friction between layers within a viscous fluid that are moving relative to each other and the surfaces slides on the other past.
- *Lubricated friction* is a case of fluid friction where a fluid separates two solid surface.
- *Skin friction* is a component of drag, the force resisting the motion of a solid body through a fluid.
- *Internal friction* is the force resisting motion between the elements making up a solid material while it undergoes deformation when the body slides each other surfaces.

When surfaces in contact move relative to each other, the friction between the two surfaces converts kinetic energy into heat. The use of friction between pieces of wood to start a fire.

The consequence of many types of friction can be wear, which may lead to performance degradation and/or damage to components. Friction is a component of the science of tribology.

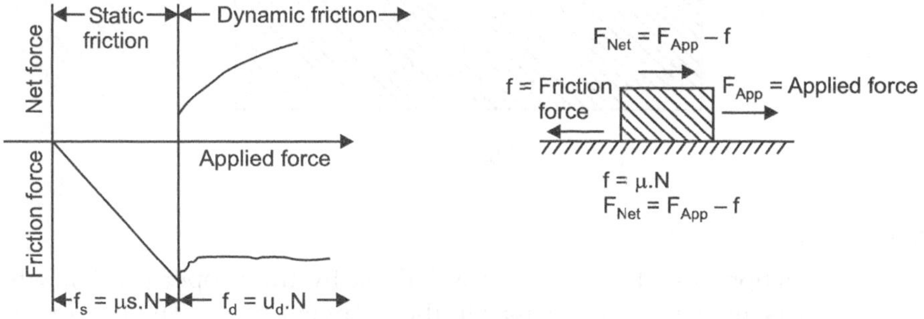


Fig. 3.1(B)

Friction is not a fundamental force but occurs because of the electromagnetic forces between charged particles which constitute the surfaces in contact. Because of the complexity of these interactions friction cannot be calculated from first principles, but instead must be found empirically as shown in Fig. 3.2 the branches of the friction.

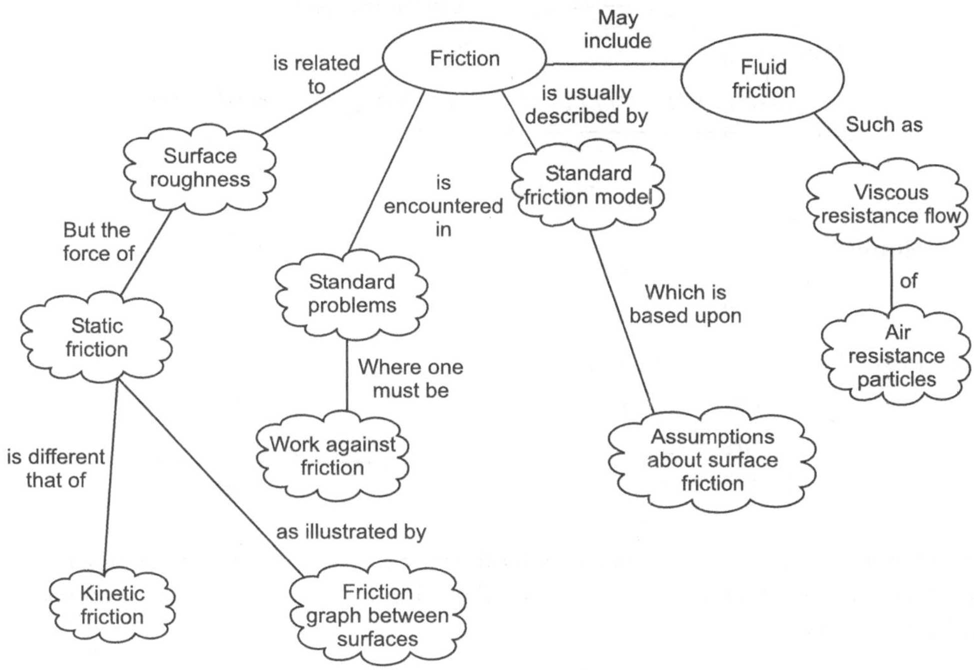


Fig. 3.2

Basic properties of friction have been described as laws:

- **Amontons' 1st Law:** The force of friction is directly proportional to the applied load.
- **Amontons' 2nd Law:** The force of friction is independent of the apparent area of contact.
- **Coulomb's Law of Friction:** Kinetic friction is independent of the sliding velocity.

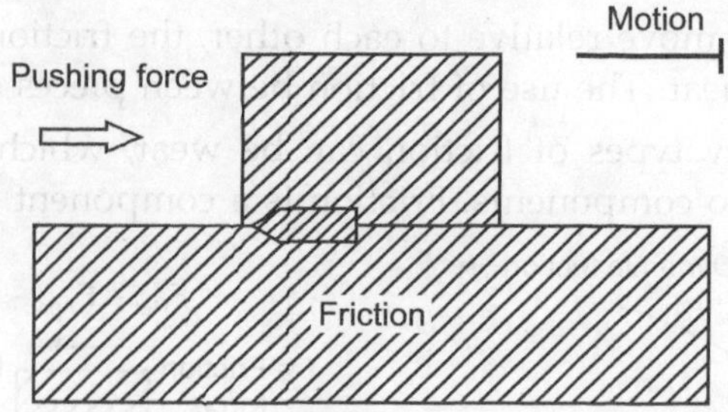


Fig. 3.3

## Law of Friction

According to law of friction smooth surfaces are defined by the properties that when they are in contact, the surfaces is always perpendicular to their tangent plane. The law of friction which has no surfaces are perfectly smooth, whenever there is a tendency for two bodies which are in contact to move relative to each other, a force known as the force of friction tends to prevent the relative motion between them. The mathematical discussion of the force of friction depends on certain assumptions which are embodied on the potential wear particles on the contact region called law of friction.

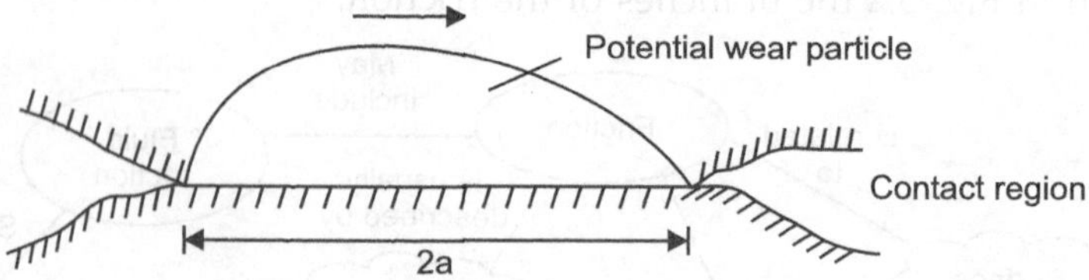


Fig. 3.4

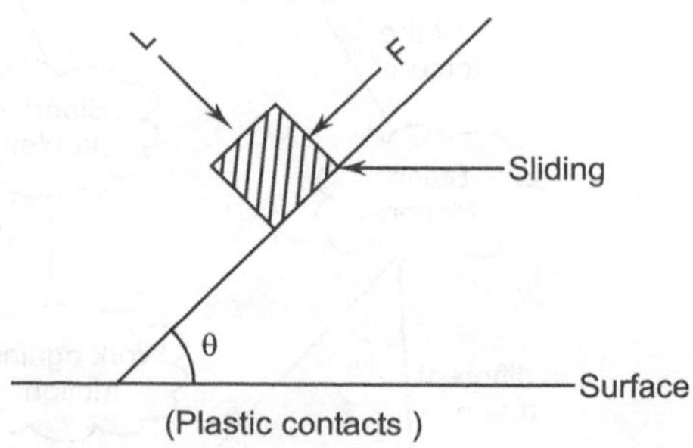


$v$ is the volume

$$v = \frac{kWx}{H}$$

$k$ is non-dimensional wear coefficient

$$dW = \pi a^2 H$$

$$dx = 2a$$

$$v \,\alpha\, \frac{1}{3}\frac{Wx}{H} = \frac{kWx}{H}$$

**Law 1:** When two bodies are in contact each other the direction of the forces of friction on one of them at it's point of contact, is opposite to the direction in which the point of contact tends to move relative to the other.

**Law 2:** If the bodies are in equilibrium, the force of friction is just sufficient to prevent friction and may therefore be determined by applying the conditions of equilibrium of all the forces acting on the body.

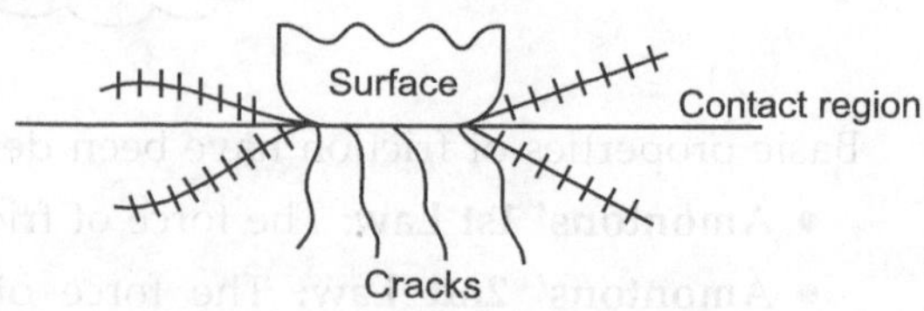

The amount of friction that can be exerted between two surfaces if the forces acting on the body are made sufficiently great, motion will occur as the friction which is exerted when equilibrium is on the point of being broken by one body sliding on another. The magnitude of limiting friction is given by the following three laws.

**Law 3:** The ratio of the limiting friction to the normal reaction between two surfaces depends on the substances of which the surfaces are composed and not on the magnitude of the normal reaction is usually denoted by $\mu$. If the normal reaction is $R$, the limiting friction is $\mu R$. For given materials polished to the same standard $\mu$ is found to be constant and independent of $R$. $\mu$ is called *the coefficient of friction.*

**Law 4:** The amount of limiting friction is independent of the area of contact between the two surfaces and of the shape of the surfaces, provided that the normal reaction is unaltered.

**Law 5:** When motion takes place the direction of friction is opposite to the direction of relative motion and independent of velocity. The magnitude of the force of friction is in a constant ratio to the normal reaction but this ratio may be slightly less than when the body is just on the point of moving.

Quantative laws are experimental basis for the mathematical treatment of friction. The force of friction is due to the non-rigidity of bodies. When one body rests on another, there is always an area of contact, which is much smaller than the apparent area and also depends on the normal pressure between the bodies. Friction is considered to be due to the fusion of materials (of which the bodies are composed) over the area of contact. Therefore friction would be proportional to the area of contact and therefore proportional to the normal pressure as assumed in the above laws.

## 3.2 THEORIES OF FRICTION

### 1. Interlocking Theory – Static and Dynamic Friction

The mathematical approach to tribology was undertaken on the surfaces with a geometrical resistance theory of "dry" friction – *the Interlocking Asperity Theory*. Euler's theory provides two terms for static and dynamic friction. The static friction coefficient is provided by the tangent of the asperity angle, *i.e.*, $\theta = \tan\theta$, while the dynamic friction coefficient is reduced by the kinetic term $2s/gt^2 \cos\theta$.

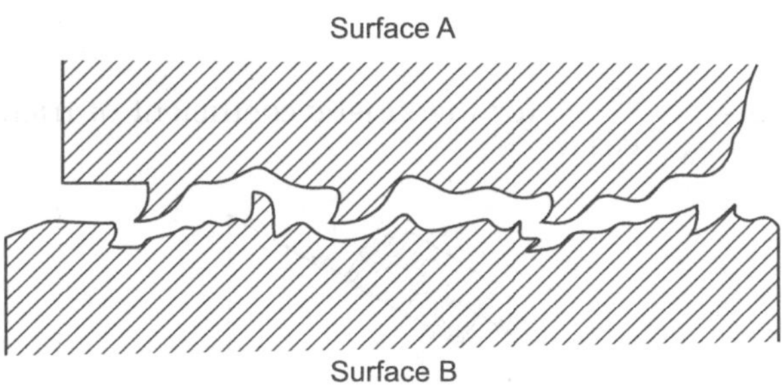


As shown in Fig. 3.5 (A) and 3.5 (B) the interlocking zone at the asperities.

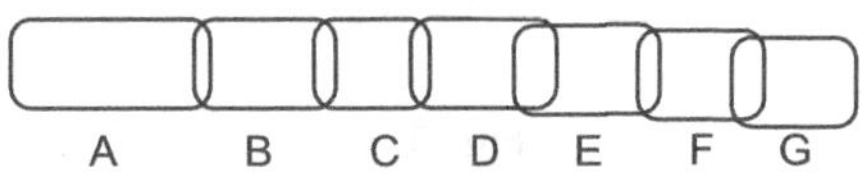


**Fig. 3.5 (A)**

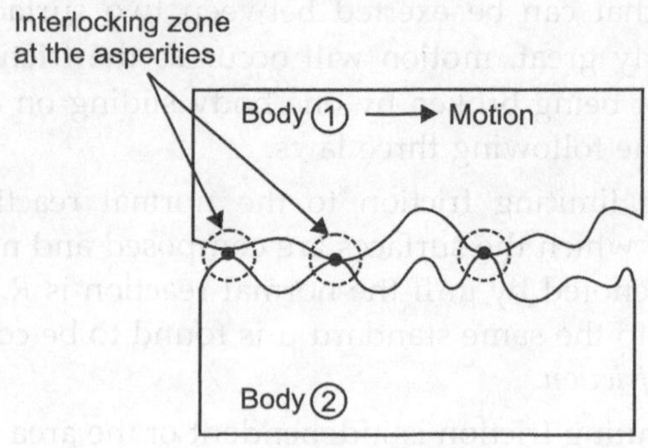


Fig. 3.5 (B)

The static equilibrium conditions of a cuboid in contact with an incline plane, Fig. 3.6. The normal force $\mathbf{F_N}$, the friction force $\mathbf{F_R}$ are given by the slope $\alpha$ and the weight $\mathbf{F}$ by:

$$F_N = F \cos \alpha$$

$$F_R = F \sin \alpha - F_S$$

with the parallel sliding force $F_S$.

The coefficient of friction $\mu$ is the quotient between friction force and normal force and therefore:

$$\mu = \frac{F_R}{F_N} = \tan \alpha - \frac{F_S}{F \cos \alpha}; \ \alpha \in \left(0, \frac{\pi}{2}\right)$$

Its represents the static friction as long as $F_S = 0$ and the kinetic friction when $F_S \neq 0$. It also results that the kinetic friction is smaller than the static friction so that the analysis is based on a strongly geometrical point of view and does not consider any forces in between the contact zone nor the real area of contact and any elastic behaviour of the two bodies. Euler fell back on the *incline plane* experiment of Galileo and set

$$F_S = \frac{2s}{gt^2} F$$

and therefore $\mu = \tan \alpha - \dfrac{2s}{gt^2 \cos \alpha}$ for the kinetic coefficient of friction.

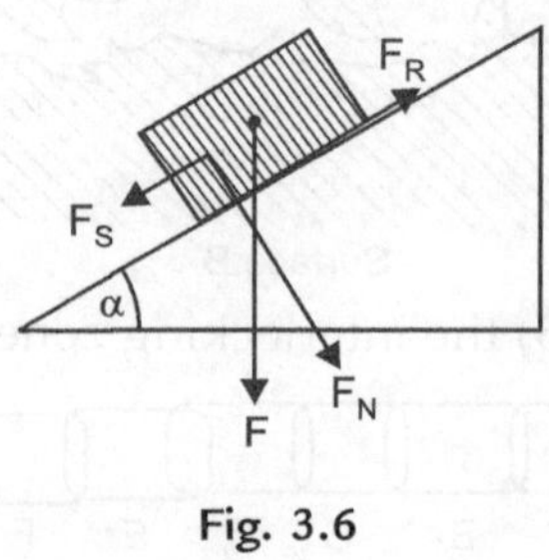


Fig. 3.6

**2. Rolling Friction**

Rolling friction is caused primarily by the interference of small indentations on surface rolls over another. The frictional forces involved with wheels, cylinders, and spheres. In the ideal case of the wheel, look at the forces acting on the wheel. In pure rolling motion friction is required to start, stop, change the motion of a wheel. We can see the frictional force necessary to begin motion, and get the wheel moving at a velocity *v*.

(*a*) Those which arise through micro-slip and friction at the contact interface of the surfaces.

(*b*) Those which are due to inelastic properties of the material, and the surface body.

(*c*) Those due to roughness of the rolling surfaces and sliding surfaces.

The frictional resistance of a rolling wheel or cylinder is proportional to the load *P*, and inversely proportional to the radius of the wheel. Coulomb's description a simple torque balance between the interfacial rolling resistance and the material resisting torque leads to an inverse square root dependence of friction in *R*. (*The rolling resistance is comprised of the product between friction force, F, and cylinder radius, R. The resisting torque is considering the applied load, P, with its lever arm θ, in which the asymmetric compliance of the material is reflected*).

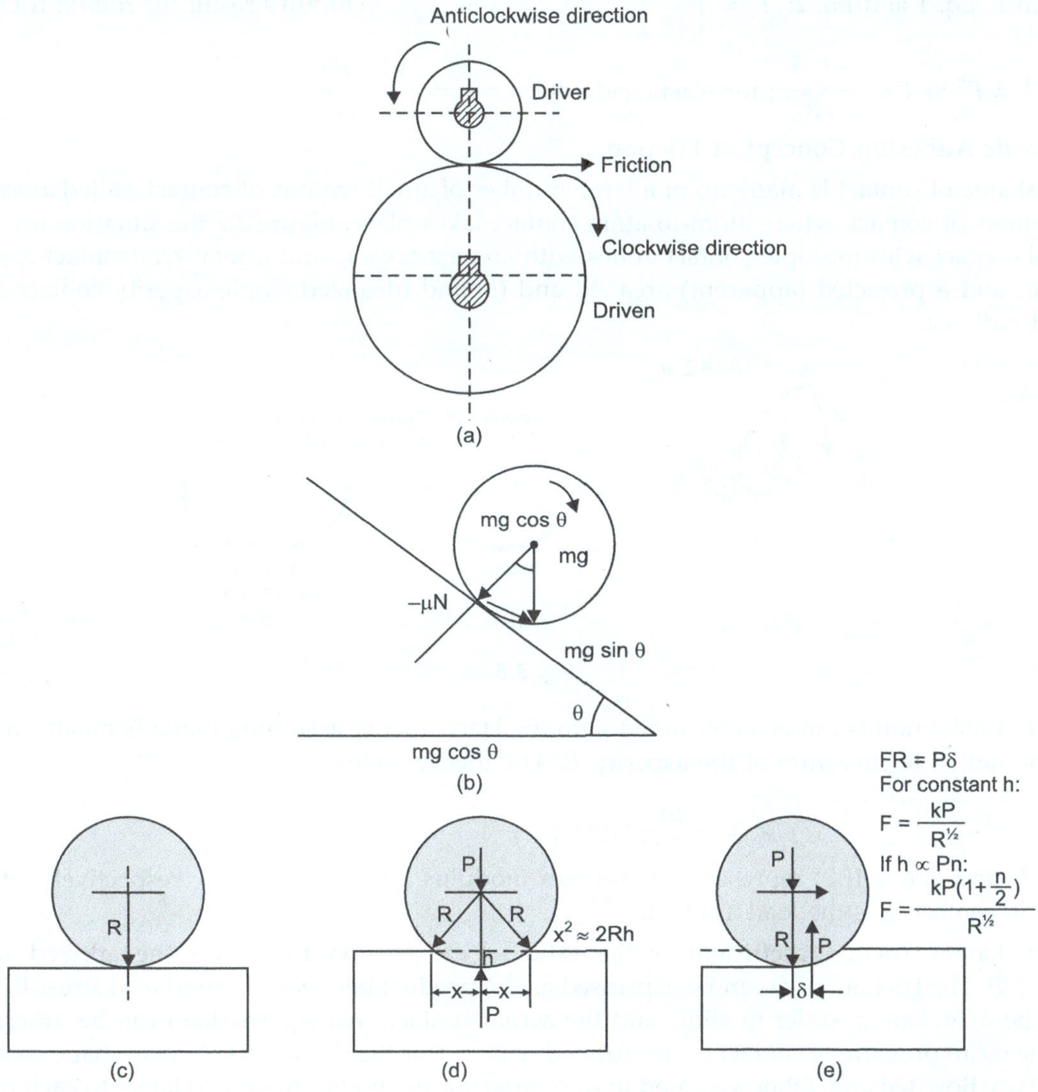


Fig. 3.7

Rolling resistance: $F \times R$ (F... Friction, R... Radius of cylinder)
Resisting torque: $P \times \delta$ (P... Load, δ... "lever arm")

Motion: $F \times R = P \times \delta \Rightarrow F = \delta\left[\frac{P}{R}\right]$ Eq. 1

From Fig. 3.7 (*b*) : $X^2 \approx 2Rh$ (X... width of half contact zone)
(h... penetration depth)

Dupuit's assumption 1: $\delta \propto X \propto \sqrt{Rh}$ Eq. 2
Dupuit's assumption 2: $h$ = constant (Where δ... lever arm)

According to (Fig. 3.7) the driver moves in the anticlockwise direction and driven moves on the clockwise direction so the friction occurred during the movement/rolling each other driver and driven both.

Hence, Eq. 1 and Eq. 2: $F \propto \left[\frac{P}{\sqrt{R}}\right]$ (Dupuit's result for rolling friction)

(if $h \propto P^n \Rightarrow F \propto \frac{P^{\left(1+\frac{n}{2}\right)}}{\sqrt{R}}$; for elastic cylinder $n \approx 1 \Rightarrow F \propto \frac{P^{3/2}}{\sqrt{R}}$

**3. Inelastic Adhesion Concept of Friction**

The real area of contact is made up of a large number of small regions of contact, called *asperities* or junctions of contact, where atom-to-atom contact takes place. Figure 3.8 the situation for (*a*) a general contact with multiple contact zones with contact area $a_i$, and a total real contact area of $A_r = \Sigma a_i$, and a projected (apparent) area $A_a$, and (*b*) and idealized single asperity contact with contact radius *a*.

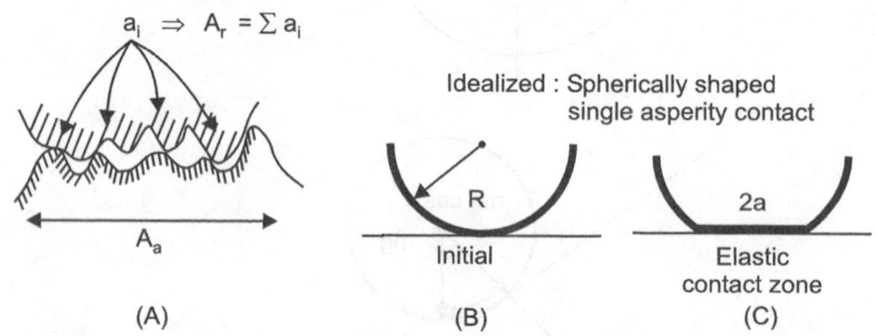


Fig. 3.8

The contact radius can be determined with the Hertz theory assuming that it is much smaller than the radius of curvature of the asperity, *R*. The theory yields: «

$$a^3 = \frac{3PR}{4E^*};\quad E^* = \left(\frac{1-v_1^2}{E_1} + \frac{1-v_2^2}{E_2}\right);\ a \ll R$$

where $E_i$ and $v_i$ ($i$ = 1, 2) represent the Young's modulus and Poisson ratio, respectively, of the bodies in contact. *P* is the load applied.

The kinetic friction coefficient is the ratio of the friction force and the applied load, *i.e.*, $\theta = F/P$. The friction force can be expressed as the product between an interfacial stress θ, that has to be overcome in order to slide, and the actual contact area $A_r$. The load can be described with the mean pressure of contact $p_m$, multiplied with $A_r$ (for the idealized spherical shape contact, $p_m = L/\theta a^2$). Bowden and Tabor assumed that in order for the bodies to slide relative to each other (*a*) the asperities are plastically deformed; *i.e.*, the mean pressure corresponds to about three times the yield pressure, $Y$, of the material, *i.e.*, $p_{m, \text{crit}} \cong 2.8\ Y$, and

(*b*) the interfacial stress component corresponds to the shear strength of the soft material $A\theta_{\text{crit}}$.

Consequently the friction coefficient can be expressed as the ratio between the shear strength of the softer material and about three times the yield pressure; *i.e.*,

$$\mu_{\text{kin}} = \frac{F}{P} = \frac{\sigma_{\text{crit}} A_r}{P_{m,\text{crit}} A_r} = \frac{\sigma_{\text{crit}}}{P_{m,\text{crit}}} \approx \frac{\sigma_{\text{crit}}}{2.8\,Y}$$

**4. Magnitude of the Dynamic Friction Coefficient**

Dynamic friction shows in most practical applications an obvious loading dependence. Recently in the boundary lubrication, friction was found to be not as dependent on load as expected. This rather unexpected finding can be understood if one considers the total stress, $\sigma$, as comprised of the intrinsic material shear strength, $\sigma_o$, plus the compression stress $\alpha p_m$; *i.e.*,

$$\sigma = \sigma_o + \alpha p_m; \quad p_m = \frac{P}{A_r}$$

The friction force $F = \alpha A_r$. For large $\alpha$ the compression term dominates, and hence, friction is clearly loading $P$ dependent with $\mu_{\text{kin}} \approx \alpha$. For small $\alpha$ values, as it is found for boundary lubricants, $\mu_{\text{kin}} \approx 0$.

$$\mu_{\text{kin}} = \frac{\Delta F^{1-2}}{\Delta P^{1-2}} = \left(\frac{(\sigma_o + \alpha p_m)\,A_r}{p_m\,A_r}\right|_1^2 = \frac{\alpha p_m}{p_m} = \alpha \Bigg) \begin{array}{l} \xrightarrow{a>>1} \mu_{kin} \approx \alpha \\ \xrightarrow{a<<1} \mu_{kin} \approx 0 \end{array}$$

**5. Elastic Model of Friction**

In model of friction assumed fully plastic deformation of the asperities, Archard suggested friction can also occur if the asperities are only elastically deformed. Shear stress is used instead of the compression stress $\alpha$ and combining them with the Hertzian formalishm for the contact radius $a$, the friction force is found to be proportional to the 2/3-power of the load $P$

$$F = \tau\pi\left(\frac{3}{4}\frac{R}{E^*}\right)^{2/3} P^{2/3}$$

## 3.3 COEFFICIENT OF FRICTION (COF)

The coefficient of friction (COF), also known as a 'friction coefficient' and symbolized by the Greek letter $\mu$, is a dimensionless scalar value which describes the ratio of the force of friction between two bodies and the force pressing them together. The coefficient of friction depends on the materials used; for example, ice on steel has a low coefficient of friction, while rubber on pavement has a high coefficient of friction. Range from near zero to greater than one — under good conditions, a tire on concrete may have a coefficient of friction when surfaces at rest relative to each other $\mu = \mu_s$, where $\mu_s$ is the *coefficient of static friction*. This is usually larger than its kinetic counterpart.

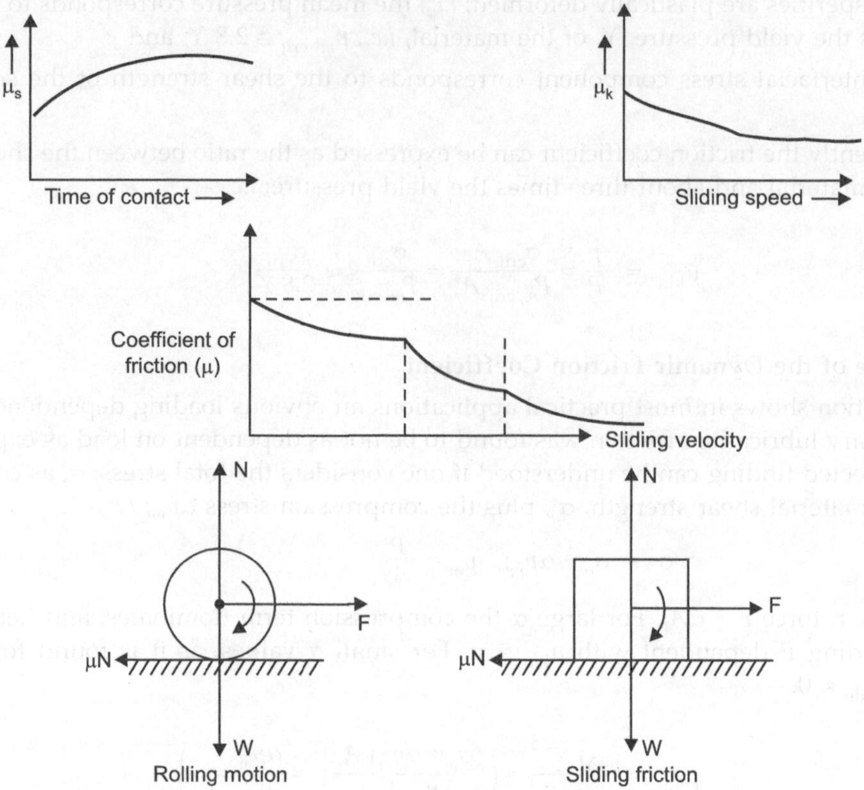


Fig. 3.9

$$\mu = \frac{F}{N} \quad \text{or} \quad F = \mu N$$

F = force required to move the body over other body

N = Normal reaction

μ = Coefficient of friction.

For surfaces in relative motion $\mu = \mu_k$, where $\mu_k$ is the *coefficient of kinetic friction*. The Coulomb's friction is equal to $F_1$, and the frictional force on each surface is exerted in the direction opposite to its motion relative to the other surface.

The coefficient of friction is an empirical measurement of rougher surfaces tend to have higher effective values. Both static and kinetic coefficients of friction depend on the pair of surfaces in contact the coefficient of static friction is *usually* larger than that of kinetic friction.

**Static Friction**

Static friction is friction between two solid objects that are not moving relative to each other. Static friction can prevent an object from sliding down a sloped surface. The coefficient of static friction, typically denoted as $\mu_s$, is usually higher than the coefficient of kinetic friction.

The static friction force must be overcome by an applied force before an object can move. The maximum possible friction force between two surfaces before sliding begins is the product of the coefficient of static friction and the normal force: $f = \mu_s F_n$. When there is no sliding occurring, the friction force can have any value from zero up to $F_{max}$. Any force smaller than $F_{max}$ attempting to slide one surface over the other is opposed by a frictional force of equal magnitude and opposite direction larger than $F_{max}$ overcomes the force of static friction and causes sliding to occur. The friction between the two surfaces is then called kinetic friction.

An example of static friction is the force that prevents a car wheel from slipping as it rolls on the ground. Even though the wheel is in motion, the patch of the in contact with the ground is stationary relative to the ground, so it is static rather than kinetic friction.

### Kinetic Friction

Kinetic (or dynamic) friction occurs when two objects are moving relative to each other and rub together. The coefficient of kinetic friction is typically denoted as $\mu_k$, and is usually less than the coefficient of static friction for the same materials how kinetic friction can be greater than static friction. Kinetic friction is now understood not to be caused by surface roughness but by chemical bonding between the surfaces. Surface roughness and contact area, however, do affect kinetic friction for micro- and nano-scale objects where surface area forces dominate inertial forces.

## 3.4 ANGLE OF FRICTION

*For the maximum angle of static friction between granular materials, see Angle of repose.*

For certain applications it is more useful to define static friction in terms of the maximum angle before which one of the items will begin sliding. This is called the *angle of friction* or *friction angle.* It is defined as:

$$\begin{aligned} \tan\theta &= \mu \\ \frac{F_L}{N} &= \mu \\ F_L &= R\sin\alpha \\ N &= R\cos\alpha \\ \frac{F_L}{N} &= \frac{R\sin\alpha}{R\cos\alpha} = \tan\alpha \end{aligned}$$

so $\mu = \tan\theta$

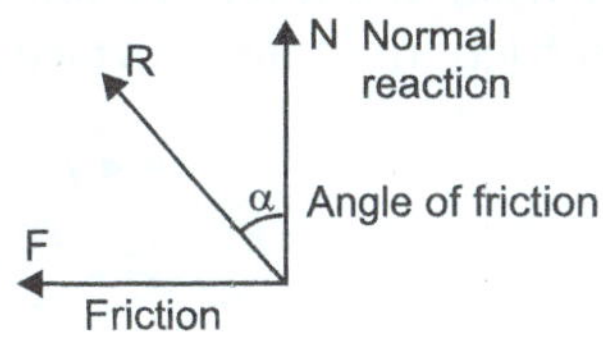


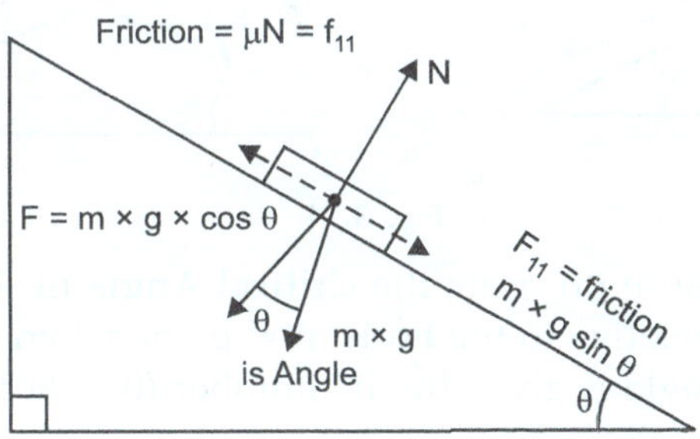


**Fig. 3.10**

where θ is the angle from horizontal and μ is the static coefficient of friction between the objects. This formula can also be used to calculate μ from empirical measurements of the friction angle.

### Energy of Friction

According to the law of conservation of energy, no energy is destroyed due to friction, though it may be lost to the system of concern. Energy is transformed from other forms into heat. A sliding hockey puck comes to rest because friction converts its kinetic energy into heat. Since heat quickly dissipates, moving objects lose energy without a driving force.

When an object is pushed along a surface, the energy converted to heat is given by:

$$E_{th} = \mu_k \int F_n(x)\,dx$$

where

$F_n$ is the normal force,

$\mu_k$ is the coefficient of kinetic friction,

$x$ is the coordinate along which the object transverses.

Energy lost to a system as a result of friction is a classic example of thermodynamic irreversibility.

**Work of Friction**

Interface between two surfaces, static friction does *no* work, because there is never displacement between the surfaces. Kinetic friction is always in the direction opposite the motion, and does *negative* work. Friction can do *positive* work in certain frames of reference. Placing a heavy box on a rug, then pulling on the rug quickly. The box slides backwards relative to the rug, but moves forward relative to the frame of reference in which the floor is stationary kinetic friction between the box and rug accelerates the box in the same direction that the box moves, doing *positive* work.

The work done by friction can translate into deformation, and heat that can affect the contact surface properties. The work of friction is used to mix and join materials such as in the process of friction welding. When mating surfaces occur when work due frictional forces rise to unacceptable levels. Harder corrosion particles caught between mating surfaces (fretting) wear of frictional forces or failure may result from excessive wear due to work of friction. Work due to friction, fit and surface finish of an object it no longer functions properly.

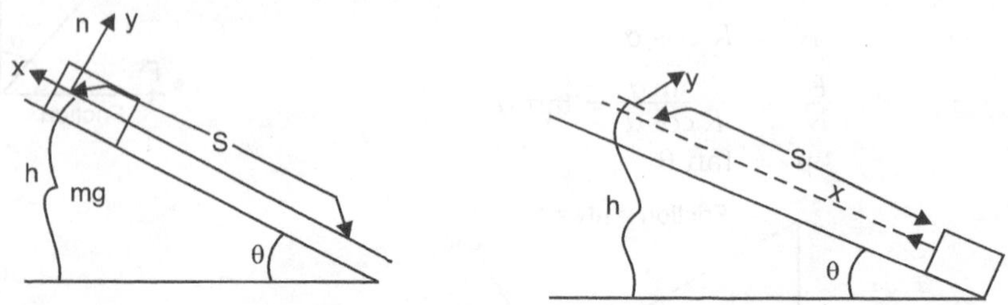


**Fig. 3.11**

The **Angle of Repose** or more precisely the **Critical Angle of Repose**, is the steepest angle of descent or dip of the slope relative to the horizontal plane when material on the slope face is on the verge of sliding. This angle is given by the number (0° – 90°) the critical deviation of an angle.

When bulk granular materials are poured onto a horizontal surface, a conical pile will form. The internal angle between the surface of the pile and the horizontal surface is known as the angle of repose and is related to the density, surface area and shapes of the particles, and the coefficient of friction of the material. Material with a low angle of repose forms flatter piles a high angle of repose is the angle between the surface of a pile and the ground.

## 3.5 STICK-SLIP PHENOMENON

The **stick-slip phenomenon**, also known as the **slip-stick phenomenon**, is the spontaneous jerking motion that can occur while two objects are sliding over each other.

Stick-slip is caused by the surfaces alternating between sticking to each other and sliding over each other, with a corresponding change in the force of friction. The static friction coefficient between two surfaces is larger than the kinetic friction coefficient. If an applied force is large enough to overcome the static friction, then the reduction of the friction to the kinetic friction can cause a sudden jump in the velocity of the movement. The attached picture shows symbolically an example of stick-slip.

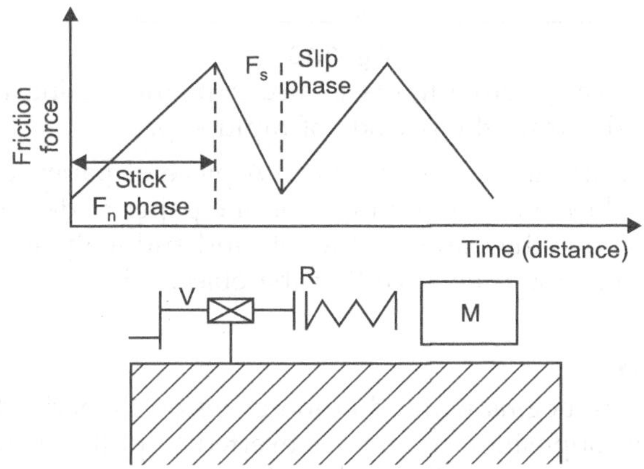


**Fig. 3.12**

*V* is a drive system, *R* is the elasticity in the system, and *M* is the load that is lying on the floor and is being pushed horizontally. When the drive system is started, the Spring *R* is loaded and its pushing force against load *M* increases until the static friction coefficient between load *M*. The load starts sliding and the friction coefficient decreases from its static value to its dynamic value. At this moment the spring can give more power and accelerates *M*. During *M*'s movement, the force of the spring decreases, until it is insufficient to overcome the dynamic friction. At this point, *M* stops. The drive system however continues, and the spring is loaded again etc.

$M$ = An element of mass

$S$ = Surface which can move at velocity $V_0$

$K$ = Spring stiffness of system

$W$ = Load on element of mass

$\mu$ = Coefficient of friction between element and surface.

Stick-slip phenomenon from hydraulic cylinders, honing machines etc. Special dopes can be added to the hydraulic fluid or the cooling fluid to overcome or minimize the stick-slip effect experienced in lathes, mill centres, and other machinery where something slides on a slideway. Another example of the stick-slip phenomenon occurs when you play musical notes with a glass harp by rubbing a wet finger along the rim of a crystal wine glass. Stick-slip can also be observed on the atomic scale using a friction force microscope. In such case, the phenomenon can be interpreted using the Tomlinson model.

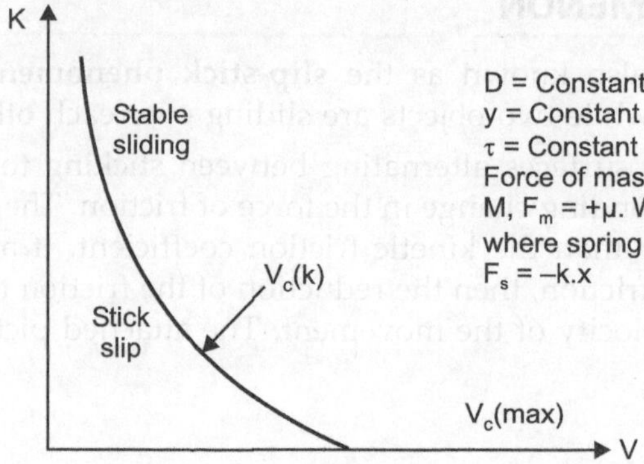


Fig. 3.13

The behaviour of seismically-active faults is also explained using a stick-slip model, with earthquakes being generated during the periods of rapid slip.

One can move objects using the stick-slip effect. A piece of paper, put the paper on a solid surface, and place a heavy object on it. Put a mark on the paper to determine the initial position of the object. Now slowly move the paper to the left and suddenly slide it to the right. If you do this a few times then you should observe that the object will be to the left side of its initial position.

**Adhesion/Cohesion Theory**

Stresses on bond performance begins with adhesion/cohesion theory. Adhesion is the force that holds substrates together in opposition to stresses exerted to pull the substrates apart cohesion is the attraction of particles that holds adhesive bonds.

Attractive force between dissimilar surfaces

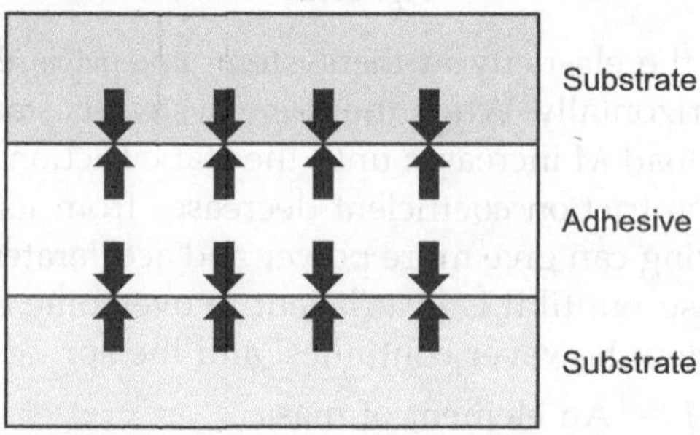


Fig. 3.14

$A \cdot P_o = W$

$A$ = Area of contact

$P_o$ = Yield pressure of metal

$W$ = Normal load.

There are three basic types of adhesion:

1. **Specific Adhesion:** The molecular attraction between contacting surfaces.
2. **Mechanical Adhesion:** Occurs when an adhesive flows into the microstructure of the surfaces to be bonded.
3. **Effective Adhesion:** Combines specific and mechanical adhesion for optimum joining strength.

Cohesion is the attraction of particles within the adhesive that combines the adhesive mass together. The combination of adhesion and cohesive strength determine bonding effectiveness.

An adhesive bond fails if either the adhesive separates adhesive breaks apart. The adhesive and cohesive strengths of some adhesives are great enough that the cohesive strength of the substrate fails before the adhesive bond.

Internal strength of material

$$F = A \,.\, St \,.\, P_c$$

$$F = A_S = \frac{W_S}{P_o}$$

$$\therefore \quad \mu = \frac{F}{W} = A_S = \frac{S}{P_o}$$

$$\mu = \frac{S_n}{P_n}$$

$$\mu \cong \frac{S_n \text{ (Soft)}}{P_o \text{ (Hard)}} = \text{low}$$

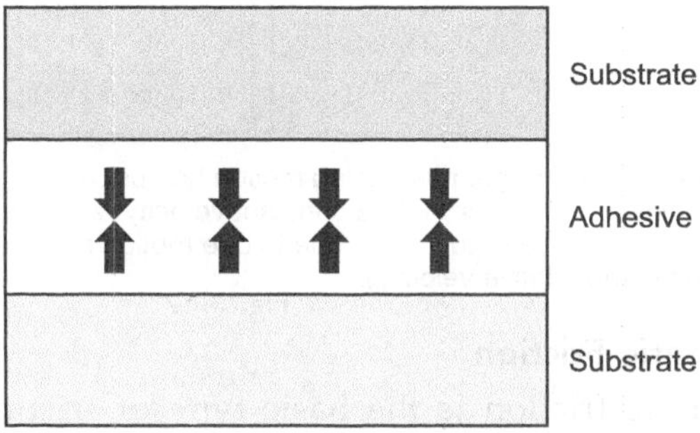


Fig. 3.15

## Rolling Friction

Friction plays a major role both in the laboratory and industrial mechanics. Friction is the resistance to the sliding, rolling, or flowing motion of an object due to its contact with another surface object. Rolling friction is caused by the interference of small indentations formed as one surface rolls over another. In pure rolling motion friction is required to start, stop, change the motion of a wheel. Below we can see the frictional force necessary to begin motion, and get the wheel moving at a velocity $v$. Friction causes the wheel to catch and stops the sliding and slipping motion; for example when a car spins its tires, slipping is taking place, thus the frictional force works to stop the spinning out and causes the tires to catch and begin pure rolling motion.

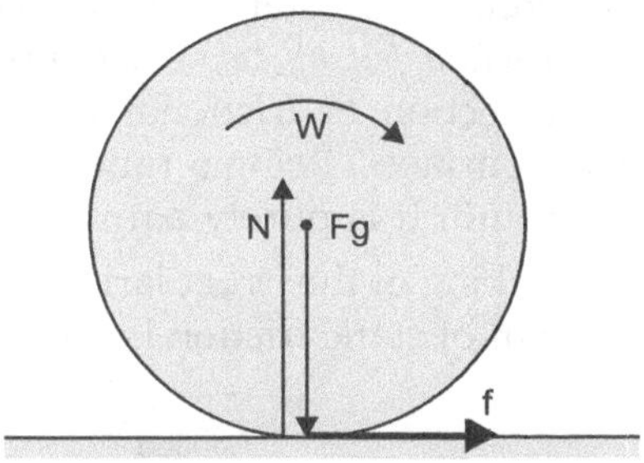


This image shows the frictional force acting on the apparent contact area in order to stop the slipping, or burning out, motion of the wheel. Notice that the wheel does not have a forward velocity as of yet the motion.

Fig. 3.16

The frictional force, $f$, the force required to slow the wheel produces a torque which tends to decrease the angular velocity, $w$. (The normal and gravitational forces produce no effect because their line of action is through the center of rotation.) Zero friction occurs only for horizontal motion at constant velocity, where acceleration is parallel to the direction of motion of centre of mass. As when the object is rolling-without-slipping up or down a sloped surface. If we consider the rotation as being about the center of mass of the object, then the frictional force must be in a direction to provide the torque necessary to decrease or increase the angular velocity, depending on whether the object is accelerating or decelerating, respectively. In pure rolling motion there is no sliding or slipping, thus the contact points have no relative motion (no relative velocity). This results in a frictional force of zero. Therefore, the wheel will roll forward with constant velocity, $v = Rw$, where $R$ is the radius of the wheel.

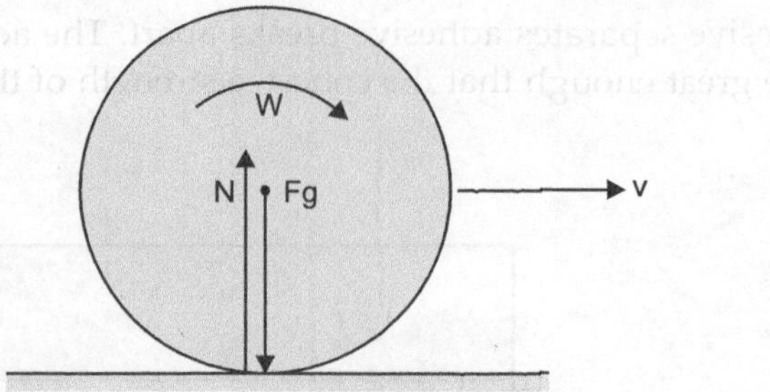


Now, in this image, pure rolling motion has been achieved and the wheel begins to roll forward with a constant velocity, v. Notice that the frictional force is now zero due to the one to one motion at the apparent contact points (no relative velocity).

**Fig. 3.17**

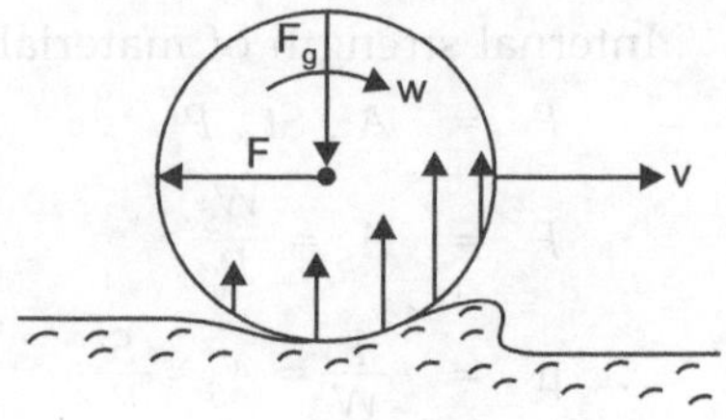


**Fig. 3.18**

## Static Friction

Static friction is the basic type of friction which opposing or resisting force between two solid surfaces, that prevents any relative motion between them. The normal forces and the frictional forces are exactly balanced motion between the two surfaces. Rock may be stationed on a slope, due to the static friction between the slope surface and the rock surface. Dry friction, one needs to understand the concept of normal force between two surfaces. It is the compressing force between two solids in contact with each other, which is perpendicular to both surfaces. For a block of wood placed on a flat surface, gravity is the normal force.

The coulomb theory describing dry friction of the static type, is governed by the following equation:

$$F_f = \mu_s F_n$$

where $F_f$ is the maximum force of friction, $F_n$ is the normal force between the surfaces in contact and $\mu_s$ is the coefficient of static friction. Thus, static friction coefficient can be defined as *'The ratio between the maximum force of friction felt by two solid surfaces in contact and the normal force exerted between two surfaces'*. The direction of frictional force is opposite to the potential movement direction between these static surfaces. Being a ratio, the static friction coefficient is a dimensionless scalar with no measuring unit. It is purely a number.

Static frictional forces from the interlocking of the irregularities of two surfaces will increase to prevent any relative motion. The coefficient of static friction is typically larger than the coefficient of kinetic friction.

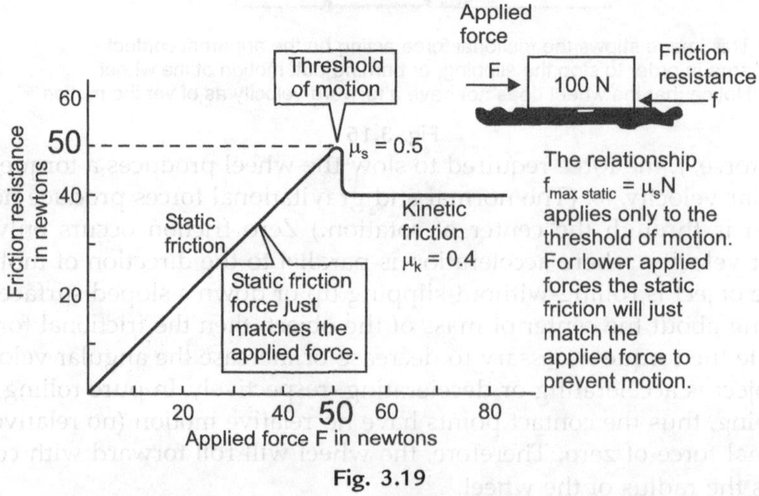


**Fig. 3.19**

The difference between static and kinetic coefficients obtained in simple experiments like wooden blocks sliding on wooden inclines roughly follows the model depicted in the friction plot shown in Fig. 3.19, 3.20 from irregularities, surface contaminants, which defy precise description. When such experiments are carried out with smooth metal blocks which are carefully cleaned, the difference between static and kinetic coefficients tends to disappear. For specific surface combinations are quoted, it is the kinetic coefficient which is generally more reliable number.

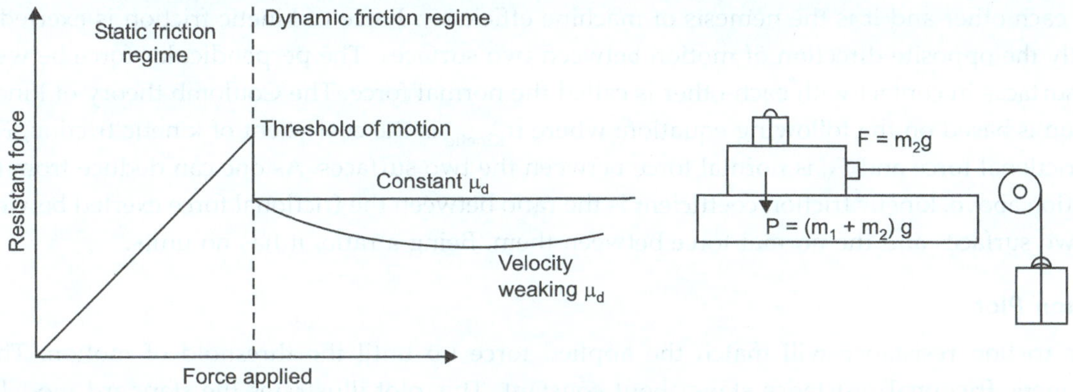


Fig. 3.20

### Kinetic Friction

Kinetic friction is the concept that emerges in empirical theory when two surfaces are moving with respect to one another, the frictional resistance is almost constant over a wide range of low speeds, model of friction the frictional force is described. The coefficient is typically less than the coefficient of static friction, reflecting the common experience that it is easier to keep something in motion across a horizontal surface than to start it in motion from rest.

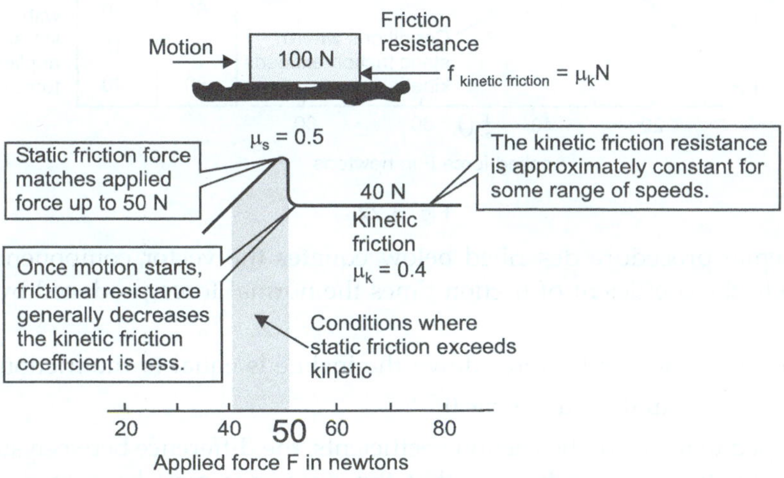


Fig. 3.21

Friction is the macroscopic force that opposes relative motion between any two surfaces at the point of contact. Entirely focuses on dry friction between solid surfaces. There are two types of dry friction. They are static and kinetic friction. Static friction occurs between two surfaces that are not in motion, whereas kinetic friction occurs between two moving surfaces in contact with each other. An example of kinetic friction is the friction experienced by car tires while running on roads. Kinetic friction is the cause of extensive wear and tear of moving surfaces in contact with each other and it is the nemesis of machine efficiency. Force of kinetic friction is exerted in exactly the opposite direction of motion between two surfaces. The perpendicular force between two surfaces in contact with each other is called the normal force. The Coulomb theory of kinetic friction is based on the following equation: where $\mu_{Kinetic}$ is the coefficient of kinetic friction, $F_f$ is the frictional force and $F_N$ is normal force between the two surfaces. As one can deduce from the equation above, kinetic friction coefficient is the ratio between the frictional force exerted between the two surfaces and the normal force between them. Being a ratio, it has no units.

## Friction Plot

Static friction resistance will match the applied force up until the threshold of motion. Then the kinetic frictional resistance stays about constant. This plot illustrates the standard model of friction.

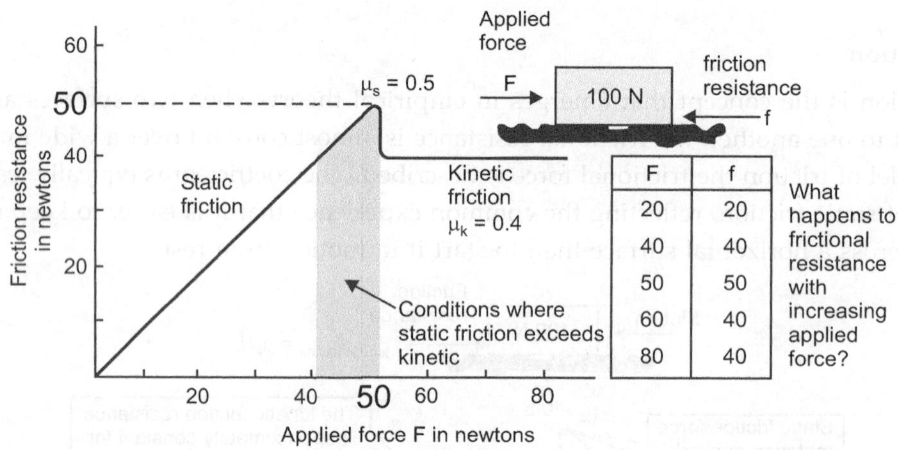


Fig. 3.22

The experimental procedure described below equates the vector component of the weight down the incline to the coefficient of friction times the normal force produced by the weight on the incline.

If the component of the gravity force down the incline is equal to the frictional force, then

$$mg \sin \theta = \mu \, mg \cos \theta$$

Surfaces are used to measure the friction coefficients, the difference between static and kinetic coefficients tends to disappear, indicating that the difference may have to do with irregular surfaces, impurities, or other factors which can be frustratingly non-reproducible. The opinion probably arises from experiences where small bits of oil or lubricant are present, or where blocks, for example, are supported by springs or other flexible supports.

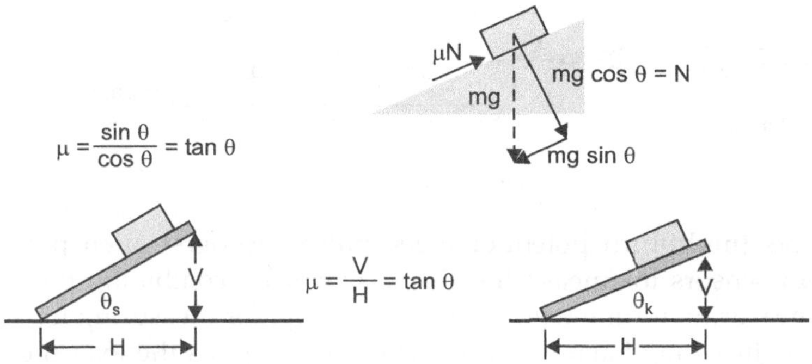


Static case: with the block at rest on the incline, raise the incline until the block starts to side. The tangent of that threshold angle is a measure of the coefficient of static friction.

Kinetic case: with the block on the incline, raise the incline in steps and bump the block gently set it into motion. If it slows to a stop, then friction overcomes gravity. Repeat to find the angle at which it moves down the incline at constant speed. The tangent of that angle is a measure of the coefficient of kinetic friction.

Fig. 3.23

## 3.6 FRICTION MEASUREMENT METHODS

### Method 1: Weight Ratio

Weight ratio method when the body and block friction by measuring the load hanging on a cord, at which the block begins to slide. The coefficient of friction is found by the quotient of the dead weight of the mass hanging on the cord and the mass of the block, *i.e.*,

$$\mu = F_f/N = m_{\text{dead weight}}/m_{\text{block}}$$

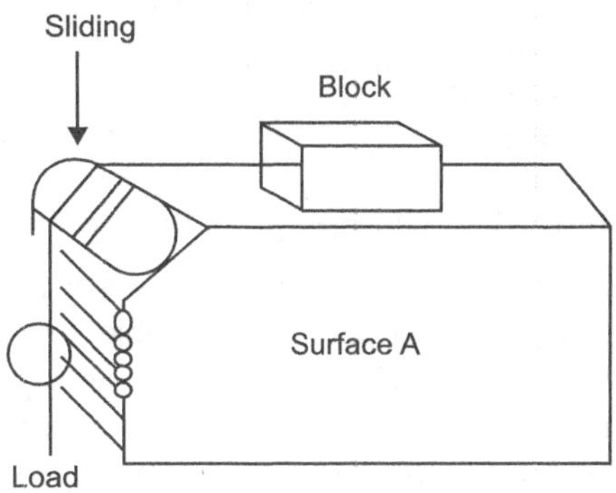


Fig. 3.24

**Static coefficient of friction – dynamic coefficient of friction:** The moment at which the block begins to slide (break away force) is the so called static friction, the force at which the block continues to slide is the dynamic or kinetic coefficient of friction. The value of the static friction exceeds that of the dynamic friction. The dynamic friction can still be dependent on velocity, contact pressure, temperature and surface roughness. The static friction can be dependent on the time that the block is in rest, which is typically the case when lubricated.

### Method 2: Spring balance

In spring balance method finding friction when pull a spring balance connected to the block and slowly increase the force until the block begins to slide. The spring balance is parallel to the surface. The reading on the spring balance scale when the load begins to slide is a measure for the static friction, while the reading when the block continues to slide is a measure of dynamic friction. The coefficient of friction is simply $\mu = F_{\text{spring}}/F_{\text{normal}} = F_{\text{spring}}/(m_{\text{block}}\, g)$, $g = 9.81$ m/s$^2$ when load "*c*" and surface "*A*" as shown in Fig. 3.25 pull the plate which is connected with the high tensied spring.

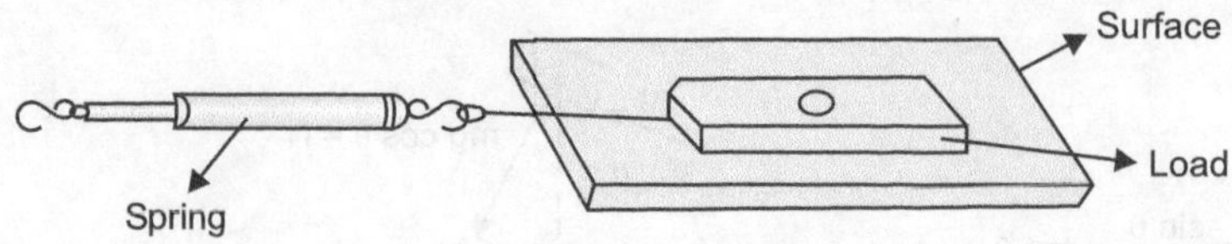


Fig. 3.25

Pulse rotation sensors (multi-turn potentiometers, pulse encoders) often prove to be very useful to create low cost sensors for measuring displacement by combining the sensor with a cable and a pulley, for measuring torque with a torsional spring, for measuring force with a wire, a pulley and a spring etc. In spring balance sensor when spring pulls the load then the transfer of load is indicating on the spring pulse rotation potentiometers and results gives in the form of pulses and finding the reading to show how much friction takes place.

### Method 3: Tilted plane

In the tilted plane, angle plays a major role where place a block on a tilted plane and increase the angle of tilt until the block begins to slide. The tangent of the tilting angle just found is the so called "friction angle". This angle is related to the coefficient of friction μ, *i.e.*, $\mu = \tan \phi = F_f/F$. This method is not suitable for continuous friction measurement over longer and stable time.

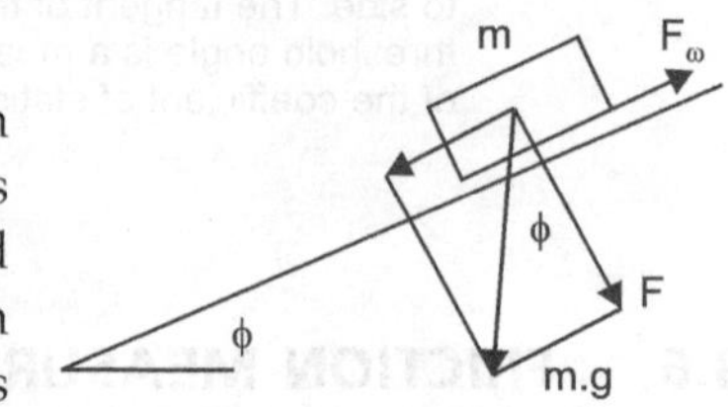


Fig. 3.26

Top of form

Angle of tilt in degrees

Angle of tilt in degrees

45

Bottom of form

### Method 4: Clamping or Pin on Cylinder

To measure the static coefficient of friction under conditions of high contact pressure the object may be clamped between two surfaces. The force necessary to put the object in motion must be halved to obtain the friction force because of the two contacting surfaces and clamping on the blocks.

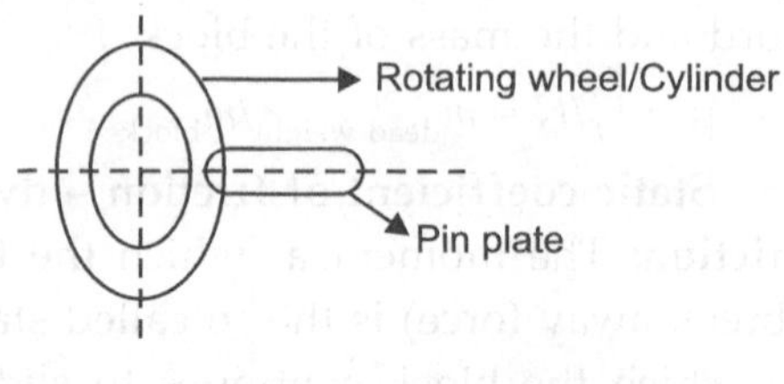


### Method 5: Pendulum

The pendulum is suitable to analyze the types of friction static and dynamic under reciprocal motion by monitoring the bearing torque. This method however requires a torque sensor. The energy loss of combined static and dynamic friction can be analyzed by considering the reduction of the amplitude of motion in time. This only requires a simple rotary potentiometer or pulse rotation sensors to visualize the amplitude reduction in time and results gives in the form of pulses according to the friction. This results finding how much friction produces in and on the surfaces of the object high pulse gives the major frequency of surface roughness of the object.

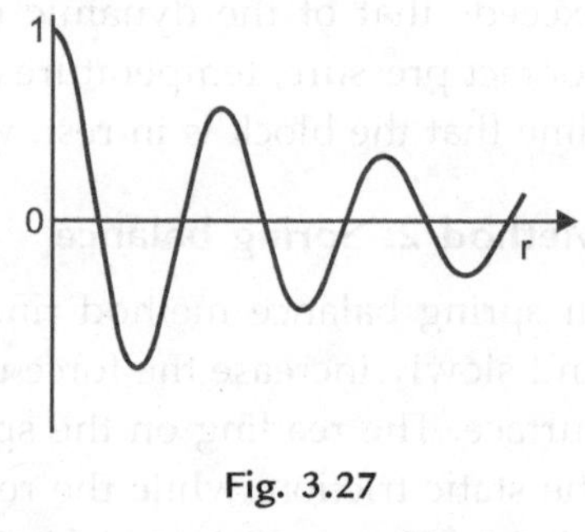


Fig. 3.27

### Method 6: Motorized Tribometers

In the measuring methods the friction coefficient is measured in fresh contacts, not after running in. The coefficient of friction may change significantly during first half hour of sliding. The time necessary to obtain a stable value of the coefficient of friction can be observed in a motorized tribometer by monitoring the friction over time. This method is common for measuring the specific wear rate and the contact temperature during operation.

## 3.7 USES OF FRICTION

Various uses of the friction to reduce or eliminate friction, it actually has some important uses.

Since friction is a resistance force that slows down or prevents motion, it is necessary in many applications to prevent slipping or sliding. In those cases, there is an advantage of having friction. However, too much friction can be a nuisance, because it can hinder motion and cause the need for expending extra energy. A good compromise is necessary to get just enough friction.

Questions you may have include:

- What are important uses of friction and friction occurs in daily life?
- How is friction a nuisance and vibratory?
- What is a good compromise for using friction in the industries?

### Important Uses

In some situations, friction is very important and beneficial. There are many things that you could not do without the force of friction.

#### Walking and Moving

In our daily life step by step friction occurs, when you walk the friction between your shoes and the ground. As you try to step forward, you push your foot backward. Friction holds your shoe to the ground, allowing you to walk. Consider how difficult it is to walk on slippery ice, where there is little friction.

#### Writing and Sliding

Writing with a pencil requires friction. You could not hold a pencil in your hand without friction. It would slip out when you tried to hold it to write. The graphite pencil led would not make a mark on the paper without friction.

A pencil eraser uses friction to rub off mistakes written in pencil lead. Rubbing the eraser on the lead wears out the eraser due to friction, while the particles worn off gather up the pencil lead from the paper.

#### Driving Car and Vehicles

When driving or moving car or any vehicle then movement occurs due to the friction between the tyres and road. So any movement of car of auto friction also necessary.

### Problems from Friction

Friction can cause problems or be a nuisance that you try to minimize.

#### Makes Movement Difficult for Object

Any time you want to move an object, friction can make the job more difficult. Excess friction can make it difficult to slide a box across the floor, ride a bicycle or walk through deep snow.

An automobile would not move forward very well unless its friction was not reduced. Oil is needed to lubricate the engine and allow its parts to move easily. Oil and ball bearings are also used in the wheels, so they will turn with little friction.

**Wastes Energy and Utilises the Work**

In any type of vehicle—such as a car, boat or airplane—excess friction means that extra fuel must be used to power the vehicle. In other words, fuel or energy is being wasted because of the friction.

Fluid friction or air resistance can greatly reduce the gas mileage in an automobile. Cars are streamlined to reduce friction. However, driving at highway speeds with your windows open can create enough drag on the car to greatly lower your gas mileage.

**Heats Parts**

The Law of Conservation of Energy states that the amount of energy remains constant. Thus, the energy that is "lost" to friction in trying to move an object is really turned to heat energy. The friction of parts rubbing together creates heat.

You've seen how people will try to start a fire by vigorously rubbing two sticks together. Or perhaps you've seen an automobile spin its wheels so much that the tires start to smoke. These are examples of friction creating heat energy. Just rub your hands together to create the same effect.

Besides the problem of losing energy to heat, there is also the threat of a part overheating due to friction. This can cause damage to a machine.

**Wears Things Out and In**

Moving parts can wear out rapidly due to friction. Lubrication is used not only to allow parts to move easier but also to prevent them from wearing out. Some other examples of materials wearing out due to friction include the soles of your shoes and a pencil eraser, auto mobile parts movements in the engine.

**Compromise**

A compromise is needed between too much friction and not enough.

**Reducing Friction between Surfaces**

Where you wanted to slide a heavy box across the floor. You would want to reduce the friction between the box and the floor, so that it would be easy to move. Lubrication is often a way to reduce friction. There are also sliders or rollers that can be used to reduce the friction how these factor we uses to maintained frictionless work.

**Need for Friction between Bodies**

When we want to increase the friction of shoes on the floor, so that it would be able to get good traction and be able to push effectively. Soles made of rubber material that include treads can reduce slipping when walking, running or pushing a heavy object.

**Summary**

Friction is necessary in many applications to prevent slipping or sliding. But also, it can be a nuisance because it can hinder motion. A good compromise is necessary to get just enough friction or a proper combination of frictions.

## 3.8 CAUSES OF FRICTION

The causes of the resistive force of friction are molecular adhesion, surface roughness, and the plowing effect.

Adhesion is the molecular force resulting when two materials are brought into close contact with each other. Trying to slide objects against each other requires breaking these adhesive bonds. Friction was caused by surface roughness, but recent studies have shown that it is actually a result of adhesive forces between the materials.

But surface roughness is a factor when the materials are rough enough to cause serious abrasion. This is called the sandpaper effect.

When one or more of the materials is relatively soft, much of the resistance to movement is caused by deformations of the objects or by a plowing effect.

Questions you may have include:

- How does adhesion cause friction or adhesion in friction?
- How does surface roughness cause friction on the surfaces?
- How do deformations cause friction on the surface of the object?

### Adhesion in Friction

When two objects are brought into contact, many atoms or molecules from one object are in such close proximity to those in the other object that molecular or electromagnetic forces attract the molecules of the two materials together. This force is called adhesion. Trying to slide one object across the other requires breaking these adhesive bonds. Adhesion is the essence of friction.

A water drop adhere to a window pane. The force of friction prevents this liquid from sliding down the solid material. But most cases of friction you see concern a solid object sliding or moving against another solid.

Sliding objects against each other require breaking these millions of contact points where the adhesion force takes effect, only to result in millions of new contact points of adhesion.

**Sticky Materials**

⇒ Some solid materials may have a composition that greatly increase their adhesion and make them even "sticky" to the touch. This stickiness greatly increases the friction. Rubber and adhesive tape are examples of sticky materials that have this type of friction.

**Fluids for Increasing or Decreasing**

⇒ Fluids often exhibit molecular adhesion, increasing the friction. This adhesion force is often seen in the capillary effect. This is where water will be pulled up a glass tube by the forces of molecular adhesion. That same force can slow down fluid motion.

⇒ One example is how a coin will easily slide down a ramp. But if you wet the coin, it will stay in place. That is because of the molecular friction of the fluid on the hard surfaces.

⇒ The motion of two fluids or two sections of a fluid against each other is also slowed down by the molecular attraction factor. This type of fluid friction is usually not considered as friction and is studied under the complex field of fluid dynamics.

**Surface Roughness of Materials**

All solid materials have some degree of surface roughness. If you looked at what seems to be a smooth surface under a high-powered microscope, you would see bumps, hills and valleys that could interfere with sliding motion.

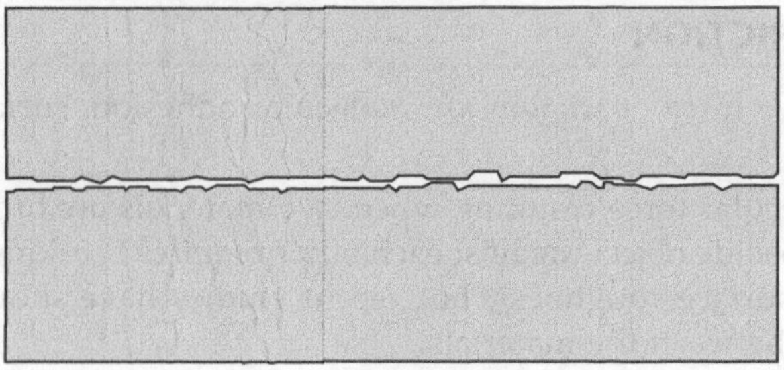

Fig. 3.28

**Close-up View of Surface Roughness**

At one time it was thought that the surface roughness of materials was the cause for friction. In reality, it only has a small effect on friction for most materials.

If the surfaces of two hard solids are extremely rough, the high points or asperities can interfere with sliding and cause friction because of the abrasion or wear that can take place when you slide one object against the other. This is the "sandpaper effect" where particles of the materials are dislodged from their surfaces. In such a case, the friction is caused by surface roughness, although the adhesion effect still plays a part in the abrasion.

**Deformations of the Particles Size**

Soft materials will deform when under pressure. This also increased the resistance to motion. For example, when you stand on a rug, you sink in slightly, which causes resistance when you try to drag your feet along the rug's surface. Another example is how rubber tires flatten out at the area on contact with the road.

When materials deform, you must "plough" through to move, thus creating a resistive force.

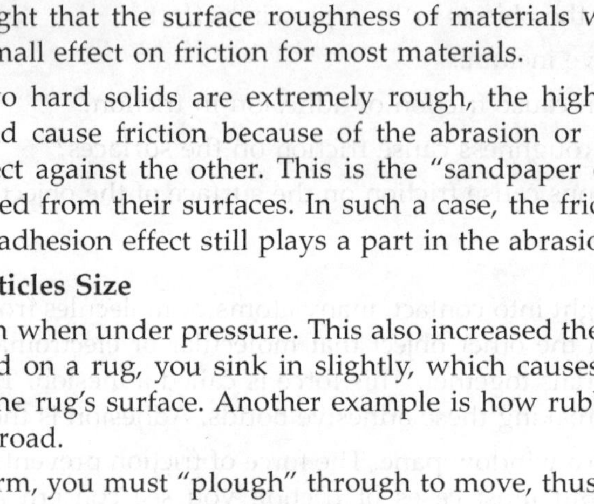

Fig. 3.29

**Pushing Object on Soft Surface**

When the deformation becomes large, such that one object sinks into the other, streamlining can affect the friction, similar to what happens in fluid friction.

**Summary**

The causes of the resistive force of friction are molecular adhesion, surface roughness, and the plowing effect. Adhesion is the molecular force resulting when two materials are brought into close contact with each other. Surface roughness is a factor in friction when the materials are rough enough to cause serious abrasion. When one or more of the materials is relatively soft, much of the resistance to movement is caused by deformations or a plowing effect.

Chapter 4

# LUBRICATION

## 4.1 INTRODUCTION

**Lubrication** is the process, or technique employed to reduce wear when surfaces moving relative to each other, by interposing a substance called lubricant between the surfaces to carry or to help carry the load (pressure generated) between the opposing surfaces. The interposed lubricant film can be a solid, (*e.g.*, graphite, $MoS_2$)a solid/liquid dispersion, a liquid, a liquid-liquid dispersion or gasses exceptionally a gas.

If applied load is carried by pressure generated within the fluid due to the frictional viscous resistance to motion of the lubricating fluid between the surfaces. To reduces wear and transmit power between various parts of product machinery.

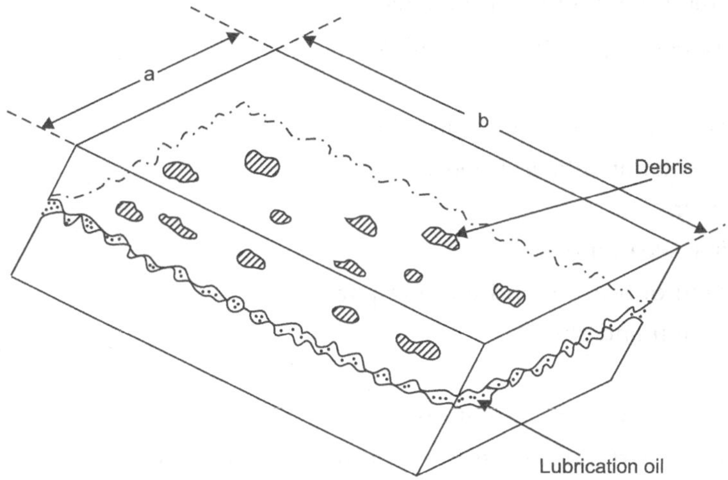


Fig. 4.1

Lubrication can also describe the phenomenon such reduction of wear occurs without human efforts. The science of friction, lubrication and wear is called Tribology. To reduces friction and

carry away contaminants and debris from the surfaces of the moving particles. Lubrication allows smooth continuous operation of equipment, with only mild wear, and without excessive stresses or seizures at bearings. When lubrication breaks down, metal or other components causing destructive damage, heat, and failure of gas, liquid and solid are interposed between two surfaces in order to improve the smoothness of movement of one surface over another and to prevent damage. When two bodies moves on each other the friction or wear occurs on the thicknesses of the surfaces then a film is generated the effectiveness of these films in preventing damage in solid contacts is commonly known as **'lubrication'**.

A lubricant is a substance (often a liquid) introduced between two moving surfaces to reduce the friction between them, improving efficiency and reducing wear. It may also have the function of dissolving or transporting foreign particles and of distributing heat. To avoid the wear and the effect of the wear on the surfaces of the bodies we generally uses, lubricants contain 90% base oil (most often petroleum fractions, called mineral oils) and less than 10% additives. Vegetable oils or synthetic liquids such as hydrogenated polyolefin, esters, silicones, fluorocarbons and many others are sometimes used as base oils. Additives deliver reduced friction and wear, increased viscosity, improved viscosity index, resistance to corrosion and oxidation, aging or contamination, etc.

Lubricants such as 2-cycle oil are added to fuels like gasoline which has low lubricity and increases the movements of transferring the power and heat between the surfaces.

Non-liquid lubricants include grease, powders (dry graphite, PTFE, Molybdenum disulfide, tungsten disulfide, etc.) Dry lubricants such as graphite, molybdenum disulfide and tungsten disulfide also offer lubrication at temperatures (up to 350°C) higher than liquid and oil-based lubricants are able to operate. On the surfaces and both the bodies which slides or rolled on each other then high or low perform to reducing friction.

## 4.2 PURPOSE/FUNCTIONS OF LUBRICANTS

Lubricants perform the following key functions:

- Keep moving parts apart from assembly
- Reduce friction from moving parts
- Transfer heat from one surface to another
- Carry away contaminants and debris
- Transmit power
- Protect against wear and friction
- Proper function of bearing and moving parts
- Prevent corrosion and slurry
- Seal for gases
- Stop the risk of smoke and fire of objects

**Keep moving parts apart:** Lubricants are typically used to separate moving parts in a system. Reducing friction and surface fatigue together with reduced heat generation, operating noise and vibrations. A thin layer of lubricant separates the moving parts. This is termed hydrodynamic lubrication. In case of high surface pressure or temperature the fluid film is much thinner and some of the forces are transmitted between the surfaces through.

**Reduce friction:** Lubricant-to-surface friction is much less than surface-to-surface friction in a system without any lubrication. Thus use of a lubricant reduces the overall system friction. Reduced friction has the benefit of reducing heat generation and reduced formation of wear particles as well as improved efficiency. Lubricants may contain additives known as friction modifiers that chemically bind to metal surfaces to reduce surface friction even when there is insufficient bulk lubricant present for hydrodynamic lubrication, *e.g.*, protecting the valve train in a car engine at startup.

**Transfer heat:** Both gas and liquid lubricants can transfer heat. However, liquid lubricants are much more effective on account of their high specific heat capacity. Lubricants may be used to warm as well as to cool when a regulated temperature is required. This circulating flow also determines the amount of heat that is carried away in any given unit of time. High flow systems can carry away a lot of heat and have the additional benefit of reducing the thermal stress on the lubricant.

An automotive oil-cooled turbocharger is a typical example. Turbochargers become red hot during operation and the oil that is cooling them only survives as its residence time in the system is very short *i.e.*, high flow rate. If the system is shut down suddenly (pulling into a service area after a high speed drive and stopping the engine) the oil that is in the turbo charger immediately oxidizes and will clog the oil ways with deposits. Non-flowing lubricants such as greases and pastes are not effective at heat transfer although they do contribute by reducing the generation of heat in the first place.

**Carry away contaminants and debris:** Lubricant circulation systems have the benefit of carrying away internally generated debris and external contaminants into the system to a filter where they can be removed. Lubricants for machines that regularly generate debris or contaminants such as automotive engines typically contain detergent and dispersant additives to assist in debris and contaminant transport to the filter and removal. In closed systems such as gear boxes the filter may be supplemented by a magnet to attract any iron fines that get created.

It is apparent that in a circulatory system the oil will only be as clean as the filter can make it, standards by which consumers can readily assess the filtering ability of various automotive filters. Poor filtration significantly reduces the life of the machine (engine) as well as making the system inefficient.

**Transmit power:** Lubricants known as hydraulic fluid are used as the working fluid in hydrostatic power transmission. Hydraulic fluids comprise a large portion of all lubricants produced in the world. The automatic transmission's torque converter is another important application for power transmission with lubricants.

**Protect against wear:** Lubricants prevent wear by keeping the moving parts apart. Lubricants may also contain anti-wear or extreme pressure additives to boost their performance against wear and fatigue.

**Prevent corrosion:** Good quality lubricants are typically formulated with additives that form chemical bonds with surfaces to prevent corrosion and rust.

**Seal for gases:** Lubricants will occupy the clearance between moving parts through the capillary force, thus sealing the clearance. This effect can be used to seal pistons and shafts.

**Why Lubrication is Required?**

To reduces wear and friction between moving surfaces and the main objective of lubrication is to prevent metal-to-metal contact between the bearing rings and the rolling elements by means of a lubricant film. In this way wear and premature roller bearing fatigue are avoided. In addition, lubrication reduces the development of noise and friction, thus improving the operating characteristics of bearing. Additional functions may include protection against corrosion and heat dissipation from bearing and high performance of parts.

Usually, bearing are lubricated with grease or oil; in rare cases where very high temperatures are involved, dry lubricants are used.

**Lubrication Types**

**1. Grease Lubrication:**

Grease lubrication is used for about 90% of all roller bearings.

The main advantage of grease lubrications are:

- A very simple design.
- It enhances the sealing effect and protective effect.
- Long service life but little maintenance is required.
- With normal operating and environmental conditions, for life grease lubrication is often possible. If a bearing is heavily stressed (load, speed, temperature), suitable re-lubrication intervals must be scheduled.

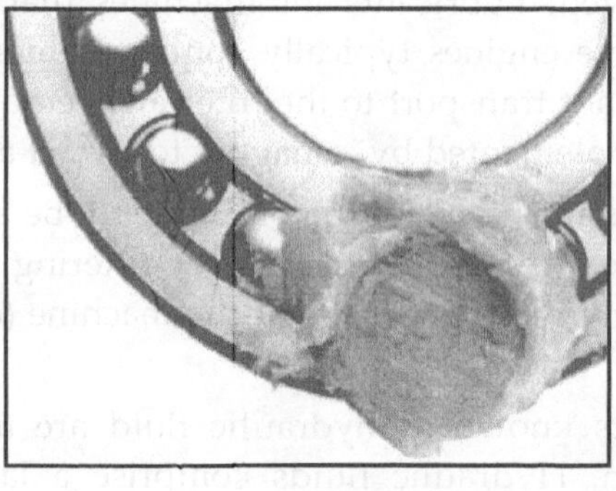

**Fig. 4.2**

**2. Oil Lubrication:**

Oil Lubrication where adjacent machine elements are already supplied with oil or where heat has to be removed by means of the lubricant. Heat can be removed by circulating substantial oil volume. It may be required where high load and/or high speed have to be accommodated or where the bearings are exposed to external heating. With oil throwaway lubrication *e.g.*, Oil-mistlubrication or oil-air-lubrication, the bearing friction is kept low. Oil lubrication mainly dependent on the thickness and viscosity of the lubricant.

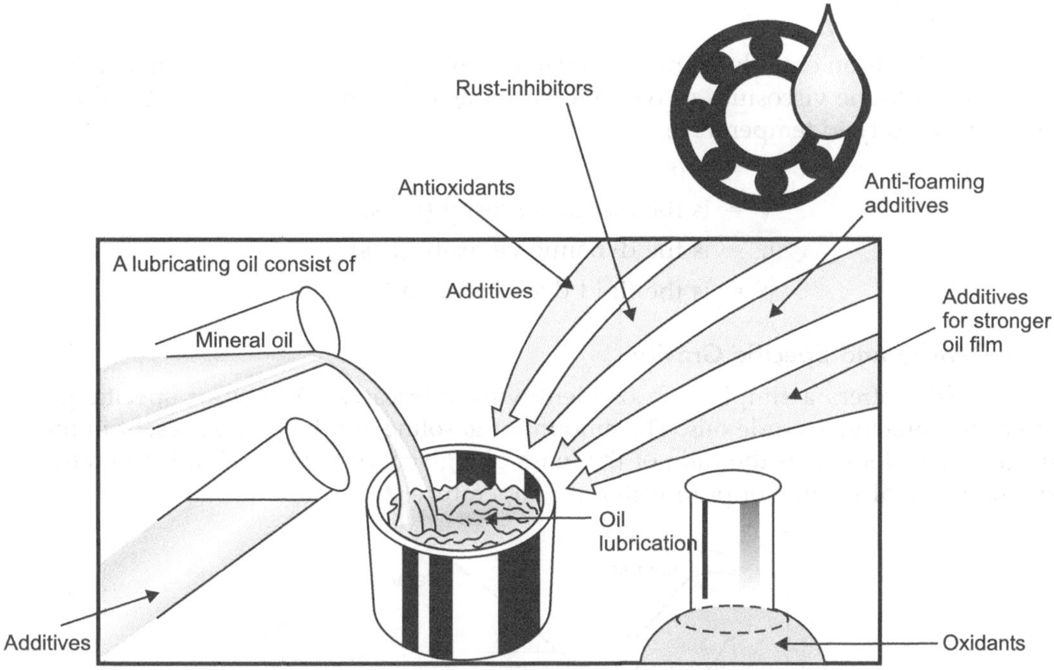


Fig. 4.3

## 4.3 PROPERTIES OF LUBRICANTS

The Properties of Lubricants know the answers to all of these Questions? What are lubricants made of, and what are their properties? Are oils different from greases? Can mineral oils be used in high performance engines? Which oils are the most suitable for application to gears, bearings, etc.? What is the oil viscosity, viscosity index, pressure-viscosity coefficient? How can these parameters be determined? What are the thermal properties and temperature characteristics of lubricants?

**Physical Properties of Lubricants:** The fundamental physical properties of lubricants such as viscosity, viscosity temperature dependence, viscosity index, pour point, flash point, volatility, oxidation stability, thermal stability, dimensions, shape and size etc.

### Oil Viscosity

The parameter which plays a fundamental role in lubrication is oil viscosity. Different oils exhibit film thickness of different viscosities. Oil viscosity changes with temperature, shear rate and pressure and the thickness of the generated oil film is usually proportional to it. Since the generated films would be thicker and a better separation of the two surfaces in contact would be achieved.

$$F \propto A \times \mu$$

where

$$F = \eta \times A \times \mu/h$$

$$\eta = (F/A)/(\mu/h)$$

The viscosity of lubricating oils is extremely sensitive to the operating temperature. With increasing temperature the viscosity of oils falls quite rapidly. In some cases the viscosity of oil can fall by about 80% with a temperature increase of 25°C. Value at the operating temperature since it determines the lubricant film thickness separating two surfaces.

### Viscosity Index

The viscosity index is an entirely empirical parameter which compares the kinematic viscosity of the oil of interest to the viscosities of two reference oils which have a considerable difference in sensitivity of viscosity to temperature.

$$\nu = \eta/\rho$$

where $\nu$ = is the kinetic viscosity ($m^2/s$)

$\eta$ = is the dynamic viscosity (Pas)

$\rho$ = is the fluid density ($kg/m^2$)

### Lubricant Density and Specific Gravity

Lubricant density offers a simple way of identifying lubricants. Density or specific gravity is often used to characterize crude oils. The amount of gasoline and kerosene present in the crude. Specific gravity is defined as the ratio of the mass of a given volume of oil at temperature '$t_1$' to the mass of an equal volume of pure water at temperature '$t_2$'.

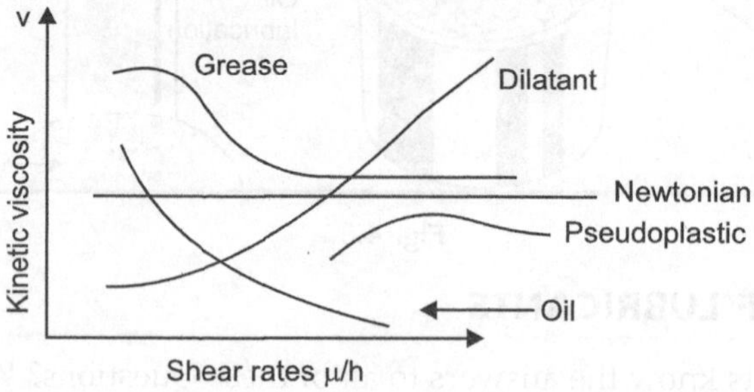


Fig. 4.4

### Thermal Properties of Lubricants

The most important thermal properties of lubricants are specific heat, thermal conductivity and thermal diffusivity. These three parameters are important in assessing the heating effects in lubrication, *i.e.*, the cooling properties of the oil, the operating temperature of the surfaces.

#### *Specific Heat*

The specific heat of an oil is usually half that of water. Synthetic hydrocarbon based lubricants, specific heat is in the range from about 1800 J/kgK at 0°C to about 3300 J/kgK at 400°C.

$$\sigma = (1.63 + 0.0034\,\theta)/S^{0.5}$$

where $\sigma$ is specific heat (kg/k)

$\theta$ is temperature of interest (°C)

$S$ is specific gravity (at 15.6°C)

#### *Thermal Conductivity*

Thermal conductivity also varies linearly with the temperature and is affected by polarity and hydrogen bonding of the molecules. The thermal conductivity of most of the mineral and synthetic hydrocarbon based lubricants is in the range between 0.14 W/mK at 0°C to 0.11 W/mK at 400°C.

$$k = (0.12/S) \times (1 - 1.667 \times 10^{-4}\,\theta)$$

where $k$ = Thermal conductivity (W/mk)

$\theta$ = temperature of interest (°C)

$\rho$ = is density (kg/m$^3$)

$\chi = k/\rho\sigma$

*Thermal Diffusivity*

Thermal diffusivity is the parameter describing the temperature propagation into the solids which:

$$\chi = k/\rho\sigma$$

The values of density, specific heat, Thermal conductivity and thermal diffusivity.

## Temperature Characteristics of Lubricants

The temperature range over which the lubricant can be used is of extreme importance. At high temperatures, oils decompose or degrade, while at low temperatures oils may become near solid or even freeze. Oils can be degraded by thermal decomposition and oxidation.

*Pour Point and Cloud Point*

The pour point of an oil is the lowest temperature at which the oil will just flow when it is cooled. To determine the pour point the oil is first heated to ensure solution of all ingredients and elimination of any influence of past thermal treatment. It is then cooled at a specific rate and, at decrements of 3°C, the container is tilted to check for any movement. The temperature 3°C above the point at which the oil stops moving is recorded as the pour point. The cloud point is the temperature at which paraffin wax and other materials begin to precipitate. If the cloud point of an oil is observed at a temperature higher than the pour point, the oil is said to have a **'Wax Pour Point'**. If the pour point is reached without a cloud point the oil shows a simple **'Viscosity Pour Point'**. There is also another critical temperature known as the **'Flock Point'**, which is primarily limited to refrigerator oils.

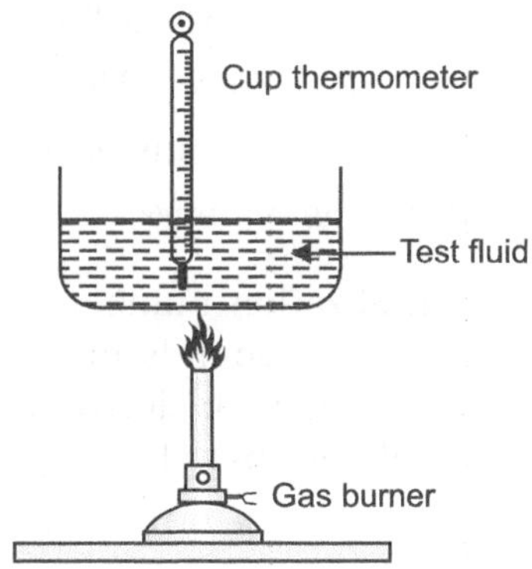


**Fig. 4.5**

*Flash Point and Fire Point*

The 'flash point' of the lubricant is the temperature at which its vapour will ignite. Flash point the oil is heated at a standard pressure to a temperature which is just high enough to produce sufficient vapour to form an ignitable mixture with air. The 'fire point' of an oil is the temperature at which enough vapour is produced to sustain burning after ignition.

*Volatility and Evaporation*

The loss of lubricant due to evaporation. The temperature has a controlling influence. At elevated temperatures in particular, oils may become more viscous and greases tend to stiffen and eventually dry out because of evaporation. Volatility of lubricants is expressed as a direct measure of evaporation losses. In order to determine the lubricant volatility, a known quantity of lubricant is exposed in a vacuum thermal balance device.

*Oxidation Stability*

Oxidation stability is the resistance of a lubricant to molecular breakdown or rearrangement at elevated temperature in the ordinary air environment. Lubricating oils can oxidize when exposed to air, particularly at elevated temperature, and this has a very strong influence on the life of the oil. The rate of oxidation depends on the degree of oil refinement, temperature, presence of metal catalysts and operating conditions. It increases with temperature. Oxidation of oils is a complex process.

*Thermal Stability*

When heated above a certain temperature oils will start to decompose, even if no oxygen is present. Thermal stability is the resistance of the lubricant to molecular breakdown or molecular rearrangement at elevated temperatures in the absence of oxygen. When heated mineral oils breakdown to methane, ethane and ethylene. Thermal stability can be improved by the refining process, but not by additives.

*Surface Tension*

Same lubricant can show different wetting and spreading characteristics depending on the degree of oxidation or on the modification of the lubricant by additives. The phenomena of wetting and spreading are dependent on surface tension.

$$\sigma_s = \frac{F}{4\pi r}$$

$\sigma_s$ = Surface Tension (N/m)

$F$ = Force [$N$]

$r$ = Radius of platinum ring

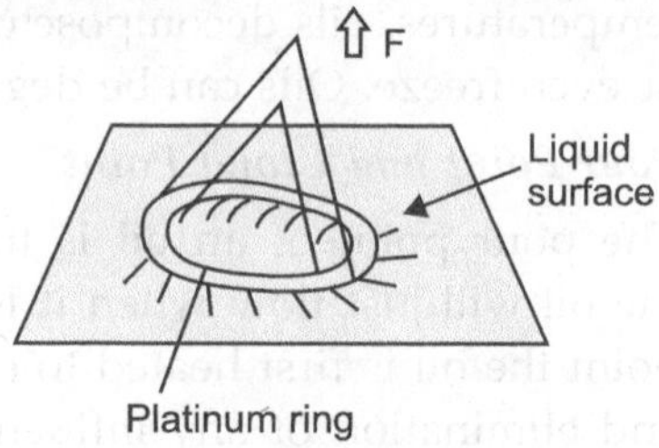


Fig. 4.6

*Neutralization Number*

The neutralization number of a lubricant is the quantity of oil necessary to neutralize acidic or alkaline compounds present in the lubricant. The procedure described in D664 is the most popular method for determining the acidic condition of the oil. The results are reported as a **Total Acid Number** (TAN) for acidic oils and as a **Total Base Number** (TBN) for alkaline oils.

where $n$ = refractive index of plate

$$\frac{(n^2 - 1)}{(n^2 + 2)} = \rho e \Rightarrow \rho = \text{oil lubricant density (g/cm}^2\text{)}$$

$e$ = constant = 0.33

## Lubricant Impurities and Contaminants

*Water Content*

Water content is the amount of water present in the lubricant. It can be expressed as parts per million, percent by volume or percent by weight. It can be measured by centrifuging, distillation and voltametry. The most popular, although least accurate, method of water content assessment is the centrifuge test. Corrosion and oxidation behaviour of lubricants is critically related to water content. An oil mixed with water gives an emulsion. An emulsion has a much lower load carrying capacity than pure oil and lubricant failure followed by damage to the operating surfaces can result. In general, in applications such as turbine oil systems.

*Sulphur Content*

Sulphur content is the amount of sulphur particles present in an oil effects on operating machinery. Sulphur is a very good boundary agent which can effectively operate under extreme conditions of pressure and temperature.

*Ash Content*

There is some quantity of incombustible material present in a lubricant measuring the amount of ash remaining after combustion of the oil. The contaminants may be wear products, solid decomposition products from a fuel or lubricant, atmospheric dust entering through a filter, etc. Some of these contaminants are removed by an oil filter but some settle into the oil.

$$C_{0.x} = 0.3 \times e^{C\ 0.63\rho\ (700 - T)\ In\ (3.33\ C\ 0.d)]}$$

$$C_0 = 7.70 \times C_{0.r} \times (980 - \rho)$$

$$\rho = \text{density of oil (kg/m}^3\text{)}$$

Excess chlorine causes corrosion whereas an insufficient amount of chlorine may cause wear and frictional losses to increase. Chlorine content can be determined either by the bomb test which provides the gravimetric evaluation, or by the volumetric test which gives chlorine content.

## Mineral Oils

Mineral oils are the most commonly used lubricants. They are manufactured from crude oil which is mined in various parts of the world. The cost of mineral oils is low and even with the rapid development of synthetic oils, solid lubricants and wear resistant polymers, their continued use in many industries seems certain.

*Manufacture of Mineral Oils*

Crude oil exhibits a complex structure which is separated into a number of fractions by a distillation process which is called fractional distillation. The process of fractional distillation involves heating the crude oil to turn it into a vapour which is then passed through a tall vertical column containing a number of trays at various levels. The vapour passes through the column and at each successive tray the temperature gradually drops. The oils presently advocated as the optimum synthetic lubricants by various oil refiners are not necessarily ideal as lubricants, but they are relatively cheap to produce and therefore are economic for large volume applications such as engine oils.

*Polyalphaolefins*

Polyalphaolefins are among the most promising general purpose synthetic lubricants. They consist of a straight carbon chain with an unsaturated carbon at one end of the chain.

*Polyphenyl Ethers*

Polyphenyl ethers exhibit better boundary characteristics than silicone oils. They have very high oxidation and thermal stability, but are limited by poor viscosity-temperature characteristics.

*Esters*

A very important group of synthetic hydrocarbons are the esters. They are produced by reacting alcohol with organic or inorganic acids. For applications such as lubrication, inorganic acids are widely used in their production.

Fatty Acid Esters have moderately low volatility, low oxidation resistance and low thermal resistance.

## Greases

Greases are not simply very viscous lubricating oils. They are in fact mixtures of lubricating oils and thickeners. A grease consists of oil constrained by minute thickener fibres. Since the oil is constrained and unable to flow it provides semi-permanent lubrication. For this reason, greases are widely used in spite of certain limitations in performance. Greases have to meet the same requirements as lubricating oils but with one extra condition, the grease must remain as a semi-solid mass in spite of high service temperatures.

*Manufacturing of Greases*

Greases are manufactured by adding alkali and fatty acid to a quantity of oil. The mixture is then heated and soap is formed from the alkali and fatty acid. Mechanical working of the grease to homogenize the composition and allow blending in of additives and the remaining oil.

*Composition*

Greases always contain three basic active ingredients: a base mineral or synthetic oil, additives and thickener. For thickeners, metal soaps and clays are used. The mineral oil plays the most important role in determining the grease performance, but in some instances the additives and the thickener can be critical.

$$\tau = \tau_p + (n_z \, du/dh)$$

**Base Oils**

Mineral oils are most often used as the base stock in grease formulation. About 99% of greases are made with mineral oils. Naphthenic oils are the most popular despite their low viscosity index. They maintain the liquid phase at low temperatures and easily combine with soaps.

**Additives**

The additives used in grease formulations are similar to those used in lubricating oils. The most common additives include anti-oxidants, rust and corrosion inhibitors, tackiness, anti-wear and extreme pressure (EP) additives. Anti-oxidants must be selected to match the individual grease. Their primary function is to protect the grease during storage and extend the service life, especially in high temperature applications.

Rust and corrosion inhibitors are added to make the grease non-corrosive to bearings operating in machinery. The function of corrosion inhibitors is to protect the non-ferrous metals against corrosion whereas the function of rust inhibitors is to protect ferrous metals.

Tackiness additives are sometimes added to impart a stringy texture and to increase the cohesion and adhesion of the grease to the surface. Anti-wear and EP additives improve, in general, the load-carrying ability in most rolling contact bearings and gears. Additives react with the surface to form protective films which prevent metal to metal contact and the consequent scoring or welding of the surfaces.

## 4.4 LUBRICANT ADDITIVES

Lubricant additives are chemicals, nearly always organic or organometallic, that are added to oils in quantities of a few weight percent to improve the lubricating capacity and durability of the oil.

**Specific purposes of lubricant additives are:**

- improving the wear and friction characteristics by provision for adsorption and E.P. lubrication,
- improving the oxidation resistance of the material surfaces.
- control of corrosion, and slurry.
- control of contamination by reaction products, wear particles and other debris,
- reducing excessive decrease of lubricant viscosity at high temperatures.
- enhancing lubricant characteristics by reducing the pour point and inhibiting the generation of foam.

The most common additives used in oil formulations contains anti-wear and E.P. lubrication additives, oxidation inhibitors, corrosion inhibitors, detergents, dispersants, viscosity improvers, pour point depressants and foam inhibitors.

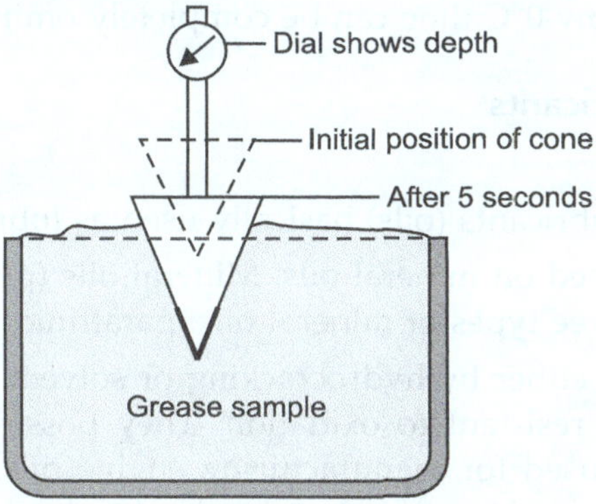


Fig. 4.7

**Wear and Friction Improvers**

Additives which improve wear and friction properties are probably the most important of all the additives used in oil formulations. Strictly speaking these chemicals are adsorption and E.P. additives and they control the lubricating performance of the oil.

**Anti-Wear Additives**

In order to protect contacting surfaces at higher temperature above the range of effectiveness of adsorption or boundary agents, anti-wear additives were designed and manufactured. There are several different types of anti-wear additives that are currently used in oil formulations.

**Extreme Pressure Additives**

These compounds are designed to react with metal surfaces under extreme conditions of load and velocity, *i.e.*, slowly moving, heavily loaded gears. E.P. additives contain usually at least one aggressive non-metal such as sulphur, antimony, iodine or chlorine. They react with exposed metallic surfaces creating protective, low shear strength surface films, which reduce friction and wear.

**Corrosion Control Additives**

Corrosion inhibitors are used for non-ferrous metals (*i.e.*, copper, aluminium, tin, cadmium, etc.) which are designed to protect their surfaces against any corrosive agents present in the oil. Rust inhibitors are needed for ferrous metals and their task is to protect ferrous surfaces against corrosion.

**Contamination Control Additives**

Engine oils are regularly exposed to fuel and combustion products which inevitably contribute to their contamination. Water also plays a major role since it accelerates the oxidation of the oil and may form an oil-water emulsion.

Additives which prevent the development of all these detrimental effects are known as **'detergents'** or **'dispersants'**. The latter term however, is more accurate. The primary functions of these additives are:

- to neutralize any acids formed during the burning of fuel, and working of engine parts,
- to prevent lacquer and varnish formation on the operating parts of the engine,
- to prevent the flocculation or agglomeration of particles and carbon deposits which may choke the oil ways.

**Pour Point Depressants**

Pour point depressants are basically the same compounds as viscosity improvers. They prevent the generation of wax crystals at low temperatures by dislocating the wax structure. The oil operating temperature will never fall below 0°C they can be completely omitted from oil formulation.

## General Classification of Lubricants

### *Mineral Lubricants*

- *Fluid lubricants* (Oils) Fluid lubricants (oils) basically used as lubricant.

Mineral fluid lubricants are based on mineral oils. Mineral oils (petroleum oils) are products of refining crude oil. There are three types of mineral oils: paraffinic, naphtenic and aromatic.

*Paraffinic oils* are produced either by hydrocracking or solvent extraction process. Paraffinic oils are relatively viscous and resistant to oxidation. They possess high flash point and high pour point. Paraffinic oils are used for manufacturing engine oils, industrial lubricants and as processing oils in rubber, textile, and paper industries. *Naphtenic oils* are produced from crude oil distillates. Most hydrocarbon molecules of naphtenic oils have saturated ring structure.

Paraffinic oils possess low viscosity, low flash point, low pour point and low resistance to oxidation. Naphtenic oils are used in moderate temperature applications, mainly for manufacturing transformer oils and metal working fluids. Hydrocarbon molecules of aromatic oils have non-saturated ring structure. Aromatic oils are dark and have high flash point. Aromatic oils are used for manufacturing seal compounds, adhesives and as plasiticezers in rubber and asphalt production.

- *Semi-fluid lubricants (greases)*

Semi-fluid lubricants are produced by emulsifying oils or fats with metallic soap and water at 218–350°C. Typical mineral oil base grease is vaseline. Grease properties are determined by a type of oil (mineral, synthetic, vegetable, animal fat), type of soap (lithium, sodium, calcium, etc. salts of long-chained fatty acids) and additives (extra pressure, corrosion protection, anti-oxidation, etc.).

Greases are used in variety applications where fluid oil is not applicable and where thick lubrication film is required: lubrication of roller bearings in railway car wheels, rolling mill bearings, steam turbines, spindles, jet engine bearings and other various machinery bearings.

- *Solid lubricants*:

Solid lubricants possess lamellar structure preventing direct contact between the sliding surfaces even at high loads. Graphite and molybdenum disulfide particles are common Solid lubricants. Boron nitride, tungsten disulfide and polytetrafluorethylene are other solid lubricants. Solid lubricants are mainly used as additives to oils and greases. Solid lubricants are also used in form of dry powder or as constituents of coatings.

### *Synthetic Lubricants*

- *Polyalphaolefins* (PAO)

Polyalphaoleins are the most popular synthetic lubticant. Chemical structure and properties are identical to those of mineral oils. Polyalphaoleins (synthetic hydrocarbons) are manufactured by polymerization of hydrocarbon molecules (alphaoleins). The process occurs in reaction of ethylene gas in presence of a metallic catalyst.

• *Polyglycols* (PAG)

Polyglycols are produced by oxidation of ethylene and propylene. The oxides are then polymerized resulting in formation of polyglycol. Polyglycols are water soluble. Polyglycols are characterized by very low coefficient of friction. They are also able to withstand high pressure without EP additives.

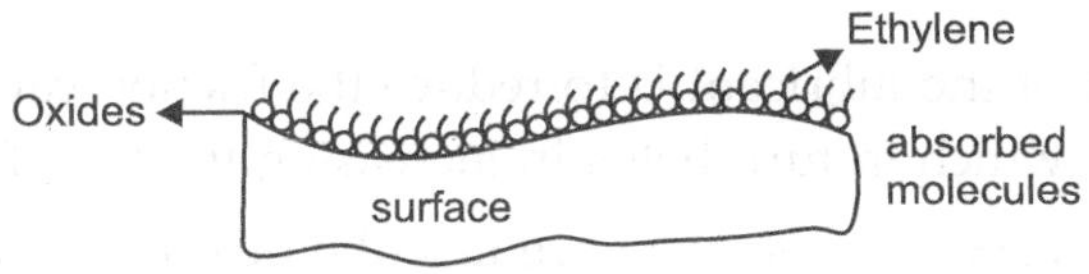


**Fig. 4.8**

• *Ester oils*

Ester oils are produced by reaction of acids and alcohols with water. Ester oils are characterized by very good high temperature and low temperature resistance.

• *Silicones*

Silicones are a group of inorganic polymers, molecules of which represent a backbone structure built from repeated chemical units (monomers) containing Si = O moieties. The most popular silicone is polydimethylsiloxane (PDMS). Its monomer is $(CH_3)_2SiO$. PDMS is produced from silicon and methylchloride. Other examples of silicones are polymethylphenylsiloxane and polydiphenylsiloxane. Viscosity of silicones depends on the length of the polymer molecules and on the degree of their cross-linking. Silicone lubricants (oils and greases) are characterized by broad temperature range: – 78°C to 320°C

***Vegetable Lubricants***

Vegetable lubricants are based on soyabean, corn, castor, canola, cotton seed and rape seed oil. Vegetable oils are environmentally friendly alternative to mineral oils since they are biodegradable. Lubrication properties of vegetable base oils are identical to those of mineral oils. The main disadvantages of vegetable lubricants are their low oxidation and temperature stabilities.

***Animal Lubricants***

Animal lubricants are produced from the animals fat. There are two main animal fats: hard fats (stearin) and soft fats (lard). Animal fats are mainly used for manufacturing greases.

Classification of lubricants by application:

- Engine oils, multifuel, Biodiesel, Biofuel.
- Gear oils, movelel oil.
- Hydraulic oils.
- Cutting fluids (coolants).
- Way lubricants/Turbine oils.
- Compressor oils, compressor fluids.
- Quenching and heat transfer oils.
- Rust protection oils/Wire rope lubricants.
- Transformer oils (insulating oils).
- Chain lubricants.

## 4.5 FUNCTIONS/OBJECTIVE OF LUBRICATION

Lubrication is used to creates difference between two meeting surfaces. Lubrication is the method or the technique to bring in the lubricant between two sliding surfaces to reduce the abrasion of the surfaces and to help in reducing the friction between the rapid movements of the engine parts.

- The basic objective of the lubricant is to reduce the friction caused between the moving parts of the engine which in turn helps in increasing the life of the engine.
- Lubricants are also used as coolants, to reduce the heat that is produced by the engine.
- Lubricants also serve as pads or cushions which minimize the noise caused by the moving engine parts and transmit power.
- Lubricants put around the engine parts also prevent the impurities from entering the engine, there by reducing the friction caused to the moving engine parts.
- Lubricants also prevent the engine components from rusting and corrosion as they do not absorb water and sliding motion.
- They also prevent the dust and debris like carbon and soot from collecting around the engine.
- Lubricants also help in chemically neutralizing the active products arising from combustion and save the properties of material.
- They also help in effectively dispelling the heat produced by the engine.
- The lubricant also acts as an adhesive between the engine's piston rings and the cylinder walls when high performation of the parts.
- The lubricant also act effectively to absorb some of the harmful by-products of combustion.
- Lubricants help in transferring heat from one part of the engine to another part of engine.
- Lubricants like the engine oil keep all the deposit forming materials in suspension and gets rid of them by oil filter, or draining out at proper intervals.
- Synthetic oils are used as lubricants in sports car and high performance cars, because they are very expensive; help in improving the over all engine performance, are low on oil consumption, maintain good engine cleanliness and help in faster starting and better pick up of the vehicle.

**Solid Lubricants/Dry Lubrication**

Graphite and molybdenum disulfide ($MoS_2$) are the predominant materials used as solid lubricant. Dry powder materials are effective lubricant additives due to their lamellar structure. The lamellas orient parallel to the surface in the direction of motion.

Particle orientation after initial sliding

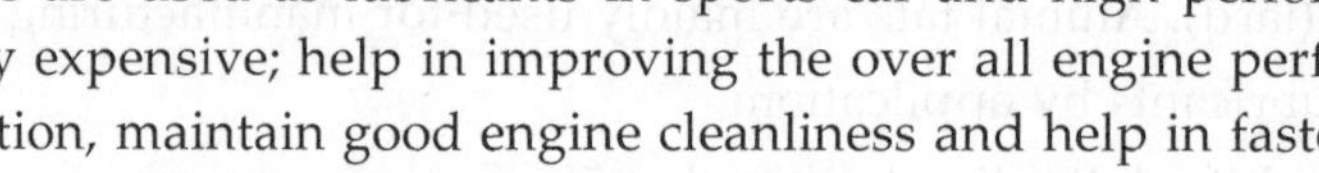

**Fig. 4.9**

The lamellas structure is able to prevent contact. In the direction of motion the lamellas easily shear over each other resulting in a low friction. Large particles best perform on relative rough surfaces at low speed, finer particle on relative smooth surface and higher speeds.

Solid lubricants include boron nitride, polytetrafluorethylene (PTFE), talc, calcium fluoride, cerium fluoride and tungsten disulfide.

- Typical applications
- Graphite
- Molybdenum Disulfide
- Boron Nitride
- PTFE
- Application Methods
- Self lubricating composites

### Typical Applications

Solid lubricants are useful for conditions when conventional lubricants are inadequate.

- *Reciprocating motion*: Sliding or reciprocating motion that requires lubrication to minimize wear as for example in gear and chain lubrication. Liquid lubricants will squeeze out while solid lubricants don't escape and prevent for fretting corrosion and galling.
- *Ceramics*: Another application is for cases where chemically active lubricant additives have not been found for a particular surface, such as polymers and ceramics.
- *High temperature*: Graphite and $MoS_2$ exhibit high temperature and oxidizing atmosphere environments, whereas liquid lubricants typically will not survive. A typical application include fasteners which are easily tightened and unscrewed after a long stay at high temperatures.
- *Extreme contact pressures*: The lamellar structure orient parallel to the sliding surface resulting in high bearing-load combined with a low shear stress. Most applications in metal forming that involve plastic deformation will utilize solid lubricants.

### Graphite

Graphite is structurally composed of planes of polycyclic carbon atoms that are hexagonal in orientation. The distance of carbon atoms between planes is longer and therefore the bonding is weaker.

Graphite is best suited for lubrication in a regular atmosphere. Water vapor is a necessary component for graphite lubrication. The adsorption of water reduces the bonding energy between the hexagonal planes of the graphite to a lower level than the adhesion energy between a substrate and the graphite. Because water vapor is a requirement for lubrication, graphite is not effective in vacuum. The thermal conductivity of graphite is generally low at 40°C.

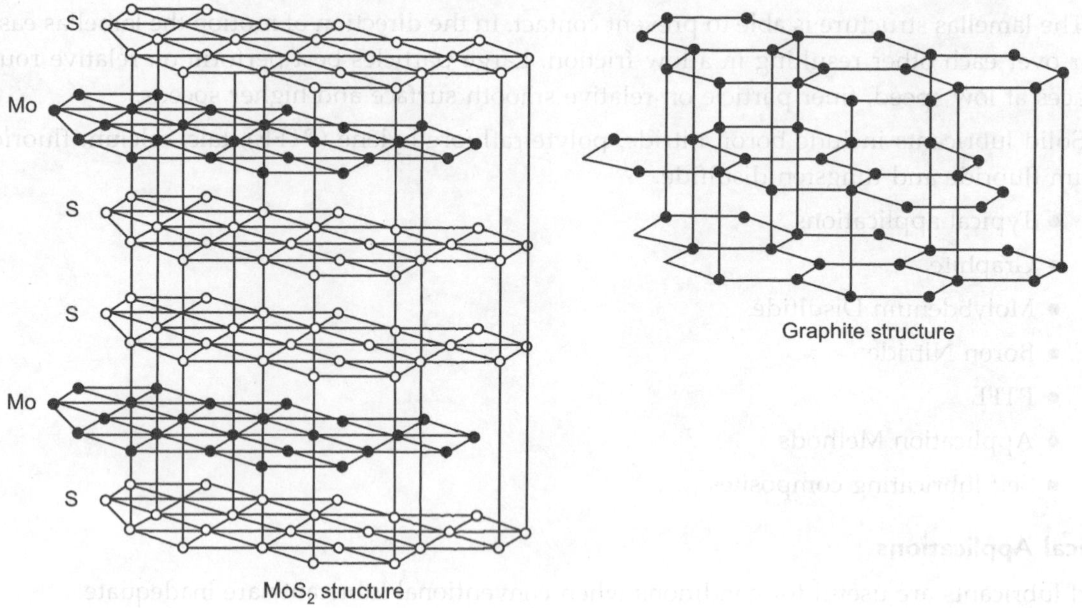


Fig. 4.10

Graphite is characterized by two main groups: natural and synthetic. Synthetic graphite is a high temperature sintered product and is characterized by its high purity of carbon (99.5–99.9%). Natural graphite is derived from mining. The quality of natural graphite varies as a result of the ore quality and post mining processing of the ore. The higher the carbon content and the degree of graphitization (high crystalline) the better the lubricity and resistance to oxidation. For applications where only a minor lubricity is needed and a more thermally insulating coating is required, then amorphous graphite would be chosen (80% carbon).

### Molybdenum Disulfide

$MoS_2$ is a mined material found in the thin veins within granite and highly refined in order to achieve a purity suitable for lubricants. Graphite has $MoS_2$ a hexagonal crystal structure with the intrinsic property of easy shear. $MoS_2$ lubrication performance often exceeds that of graphite and is effective in vacuum as well whereas graphite does not. Large particles may result in excessive wear by abrasion caused by impurities in the $MoS_2$, small particles may result in accelerated oxidation.

### Boron Nitride

Boron Nitride is a ceramic powder lubricant. The most interesting lubricant feature is its high temperature resistance of 1200°C service temperature in an oxidizing atmosphere. Further Boron has a high thermal conductivity. Boron is available in two chemical structures, *i.e.,* cubic and hexagonal where the last is the lubricating version. The cubic structure is very hard and used as an abrasive and cutting tool component.

### PTFE

PTFE is widely used as an additive in lubricating oils and greases. Due to the low surface energy of PTFE, stable unflocculated dispersions of PTFE in oil or water can be produced. PTFE does not have a layered structure. The macro molecules of PTFE slip easily along each other, similar to lamellar structures. PTFE shows one of the smallest coefficients of static and dynamic friction, down to 0.04. Operating temperatures are limited to about 260°C.

## Application Methods

*Spraying/dipping/brushing*: Dispersion of solid lubricant as an additive in oil, water or grease is most common used for lubrication after assembly a dry film lubricant can be sprayed. After the solvent evaporates, the coating cures at room temperature to form a solid lubricant. Pastes are grease like lubricants containing a high percentage of solid lubricants used for assembly and lubrication of highly loaded, slow moving parts.

Free powders has its limitations, since adhesion of the solid particles to the substrate is usually insufficient to provide any service life in continuous applications to improve running-in conditions or in metal forming processes a short duration of the improved slide conditions may suffice.

*AF-coatings*: Anti-friction coatings are "lubricating paints" consisting of fine particles of lubricating pigments, such as molydisulfide, PTFE or graphite, blended with a binder. Dry film lubricants also contain special rust inhibitors which offer exceptional corrosion protection. AF-coatings are applied where fretting and galling is a problem where operating pressures exceed the load-bearing capacities of ordinary oils and greases, where smooth running in is desired (piston, camshaft), where clean operation is desired (AF-coatings will not collect dirt and debris like greases and oils), where parts may be stored for long periods of time.

## Composites

*Self lubricating composites*: Solid lubricants as PTFE, graphite, $MoS_2$ and some other anti-friction and anti wear additives are often compounded in polymers and all kind of sintered materials. $MoS_2$ for example is compounded in materials for sleeve bearings, elastomere O-rings, carbon brushes etc. Solid lubricants are compounded in plastics to form a "Self lubricating" or "Internally lubricated" thermoplastic composite.

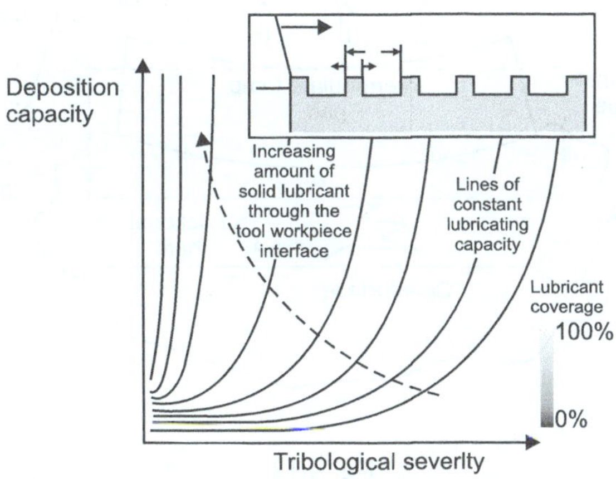


Fig. 4.11

## 4.6 ADVANTAGES OF SOLID LUBRICANTS

- Have high heat dissipation with high thermally conductive lubricants, such as diamond films used as the cutter method also.
- Have high resistance to deterioration in high-radiation environments.
- Have high resistance to abrasion in high-dust environments, access for servicing difficult.
- Have high resistance to deterioration in reactive environments conditions.
- Are more effective than fluid lubricants at intermittent loading, high loads, and high speeds performances.
- Enable equipment to be lighter and simpler because lubrication distribution systems and seals are not required.
- Can provide translucent or transparent coatings, such as diamond and diamond like carbon films, where desirable.

### Extreme Pressure Lubrication

When the moving or sliding surfaces are under very high pressure and speed, a high local temperature liquid lubricant fails to stick to the moving parts and may decompose and even vaporize. To meet this extreme pressure condition, special additives are added to the minerals oils. These are called "extreme pressure lubrication". These additives form on the metal surfaces more durable films capable of withstanding high loads and high temperature. Additives are organic compounds like chlorine (as in chlorinated esters), sulphur (as in sulphurized oils), and phosphorus (as in tricresyl phosphate).

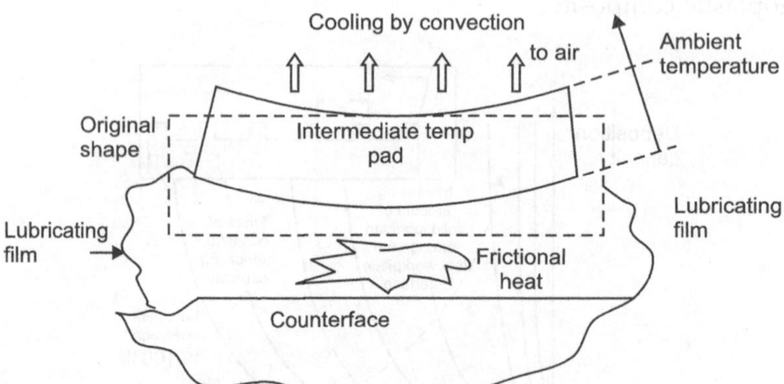


**Fig. 4.12**

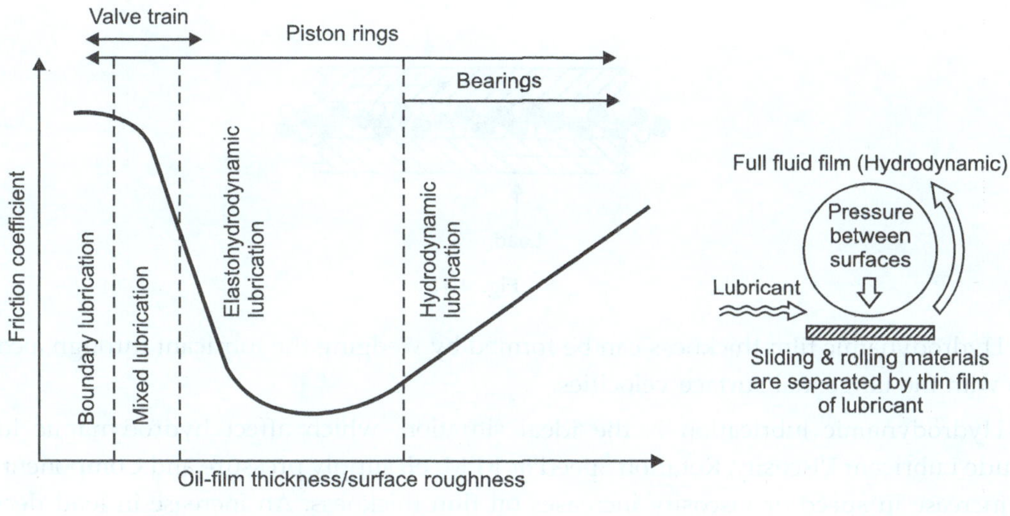


Fig. 4.13

## 4.7 HYDRODYNAMIC OR FULL FLUID FILM LUBRICATION

Lubrication by this method can be divided into four basic types or regimes and fluid action on surfaces.

When a fluid lubricant is present between two rolling and/or sliding surfaces, a thicker pressurized film can be generated by the movement of the surfaces (velocities). The non-compressible nature of this film separates the surfaces resulting in no metal to metal contact.

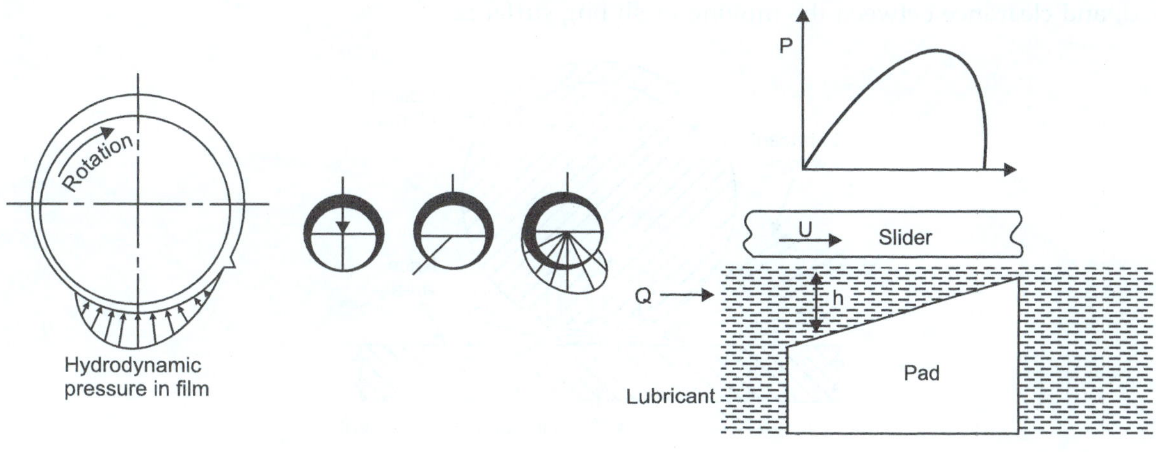


Fig. 4.14

Surfaces are completely separated by a continuous film of lubricating fluid is commonly referred to as Hydrodynamic or Full Fluid Film Lubrication. In this regime, the lubricant immediately adjacent to each surface travels at the same speed and direction of each surface. The width of the film, differential speed graduates and direction is reversed.

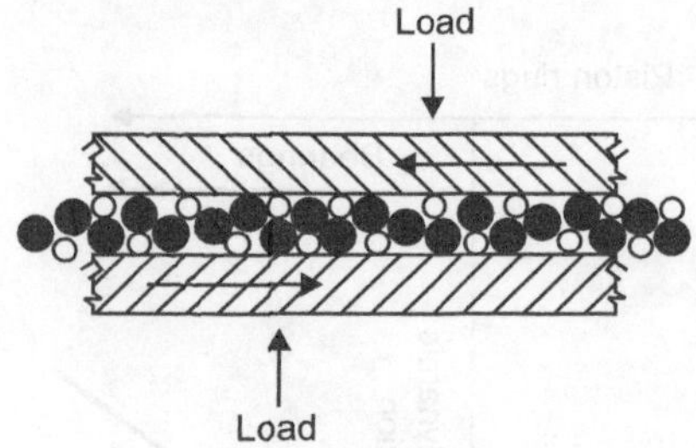


Fig. 4.15

Hydrodynamic film thickness can be formed by wedging the lubricant through a convergent gap with the tangential surface velocities.

Hydrodynamic lubrication is the ideal situation, which affect hydrodynamic lubrication include Lubricant Viscosity, Rotation Speed or RPM, oil supply pressure and Component Loading. An increase in speed or viscosity increases oil film thickness. An increase in load decreases oil film thickness.

Hydrodynamic lubrication exist when the moving surfaces are separated by the pressure of a continuous unbroken film or layer of lubrication. In this type of lubrication, the load is taken completely by the oil film.

Hydrodynamic lubrication is the formation of an oil wedge. When the journal rotates, it creates an oil taper or wedge between the two surfaces, and the pressure build up with the oil film supports the load.

Hydrodynamic lubrication depends on the relative speed between the surfaces, oil viscosity, load, and clearance between the moving or sliding surfaces.

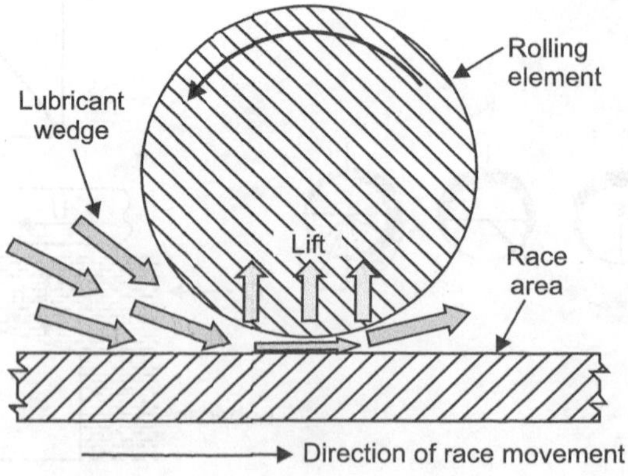


Fig. 4.16

In hydrodynamic lubrication the lube oil film thickness is greater than outlet, pressure at the inlet increases quickly, remains fairly steady having a maximum value a little to the outside of the bearing center line, and then decreases quickly to zero at the outlet.

Application of hydrodynamic lubrication.

- Delicate instruments.

- Light machines like watches, clocks, guns, sewing machines.
- Scientific instruments and motion objects.
- Large plain bearings like pedestal bearings, main bearing of diesel engines.

*Hydrocarbon oils* are considered to be satisfactory lubrication for fluid film lubrication. In order to maintain the viscosity of the oil in all seasons of the year, ordinary hydrocarbon lubricants are blended with selected long chain polymers.

## 4.8 BOUNDARY LUBRICATION

Boundary Lubrication is a condition in which the lubricant film becomes too thin to provide total separation according to Fig. 4.17 shows the motion and surfaces where load is applied and friction reduction and wear protection is then provided via chemical compounds rather than properties of the lubricating fluid. Boundary lubrication often occurs during the start up and shut down of equipment or when loading becomes excessive.

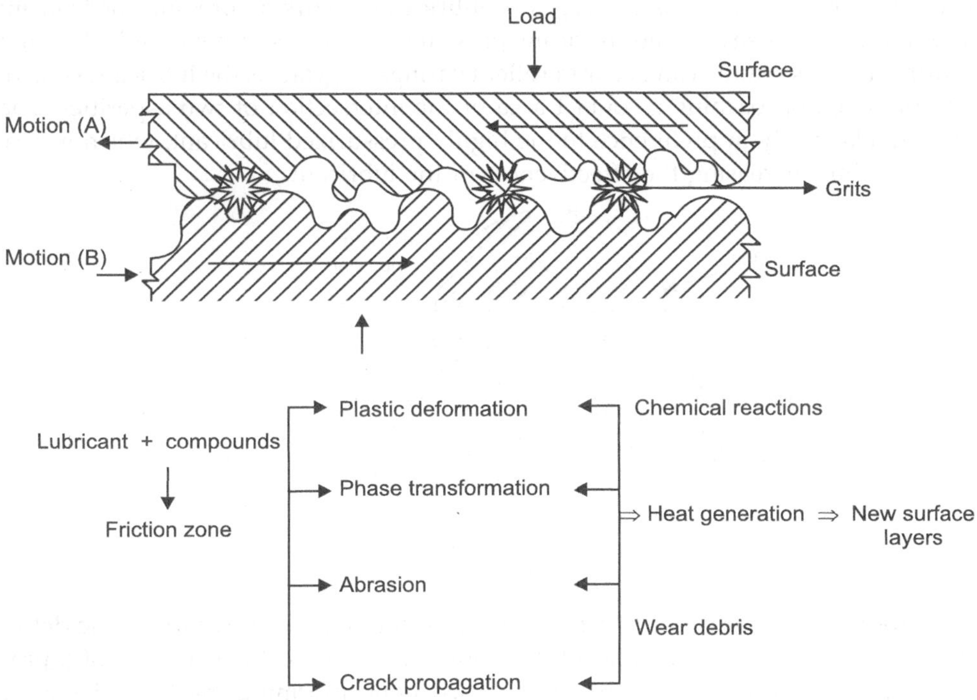


Fig. 4.17

**Mixed Film Lubrication** is a combination of both hydrodynamic and boundary lubrication. Occasional asperity contact occurs.

Boundary lubrication exists when the operating condition is not possible to establish a full fluid condition, particularly at low relative speeds between the moving or sliding surfaces.

The oil film thickness may be reduced to such a degree that metal to metal contact occurs between the moving surfaces. The oil film thickness is so small that oiliness becomes predominant for boundary lubrication.

Boundary lubrication happens when

- A shaft starts moving from rest.
- The speed is very low between the bodies.
- The load is very high on the edges of object.
- Viscosity of the lubricant is too low.

Examples for boundary lubrication:

- Guide and guide shoe in two stroke engine.
- Lubrication of the journal bearing in diesel engines (mainly during starting and stopping of engine).
- Piston rings and when cylinder liner is at TDC and BDC position when the piston direction changes and if the relative speed is very slow.

## 4.9 ELASTOHYDRODYNAMIC LUBRICATION

Elastohydrodynamic Lubrication is the types of lubricants occurs as pressure or load increases to a level where the viscosity of the lubricant provides a higher shear strength than the metal surface it supports. This regime can occur in roller bearings or gears as the lubricant is carried into the convergent zone approaching a contact area or the intersection of two asperities. The metal surfaces deform elastically in preference to the highly pressurized lubricant, which increases the contact area and thus increasing the effectiveness of the lubricant.

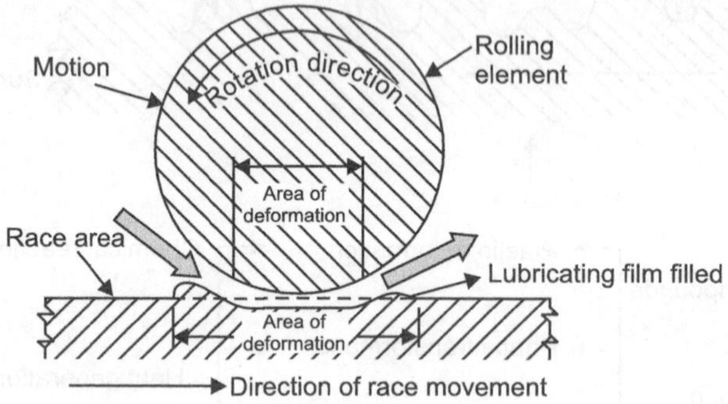


**Fig. 4.18**

Elastohydrodynamic lubrication is a form of fluid-film lubrication where elastic deformation of the bearing surfaces becomes significant. It is not only revealed the existence of a previously unsuspected regime of lubrication in highly stressed nonconforming machine elements, but it also brought order to the complete spectrum of lubrication regimes ranging from boundary to hydrodynamic.

## 4.10 HYDROSTATIC LUBRICATION

Hydrostatic lubrication is essentially a form of hydrodynamic lubrication in which the metal surfaces are separated by a complete film of oil, but instead of being self-generated, the separating pressure is supplied by an external oil pump. Hydrostatic lubrication depends on the inlet pressure of lube oil and clearance between the metal surfaces, whereas in hydrodynamic lubrication it depends on the relative speed between the surfaces, oil viscosity, load on the surfaces, and clearance between the moving surfaces.

**Example:** the cross head pin bearing or gudgeon pin bearing in two stroke engines employs this hydrostatic lubrication mechanism. In the cross head bearing, the load is very high and the motion is not continuous as the bearing oscillation is fairly short. The oil is supplied under pressure at the bottom of bearing. The lube oil pump pressure is related to the load, bearing clearance, and thickness of the oil film required, but is usually in the order of 35–140 kg/cm$^2$.

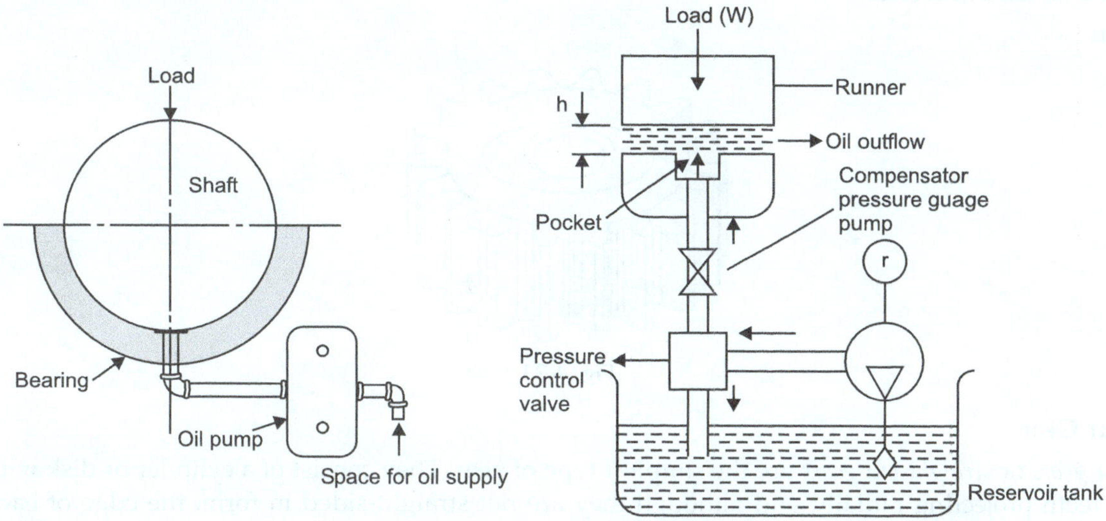


Fig. 4.19

## 4.11 GEAR

A **gear** is a rotating machine part having cut *teeth*, or *cogs*, which *mesh* with another toothed part in order to transmit torque. Two or more gears working in tandem are called a *transmission* and can produce a mechanical advantage through a gear ratio and thus may be considered a simple machine. Geared devices can change the speed, magnitude, and direction of a power source. The most common situation is for a gear to mesh with another gear, however a gear can also mesh a non-rotating toothed part, called a rack, thereby producing translation instead of rotation.

The gears in a transmission are analogous to the wheels in a pulley. An advantage of gears is that the teeth of a gear prevent slipping. When two gears of unequal number of teeth are combined a mechanical advantage is produced, with both the rotational speeds and the torques of the two gears differing in a simple relationship. In transmissions which offer multiple gear ratios, such as bicycles and cars, the term **gear**, as in *first gear*, refers to a gear ratio rather than an actual physical gear. The term is used to describe similar devices even when gear ratio is continuous rather than discrete, or when the device does not actually contain any gear, as in a continuously variable transmission.

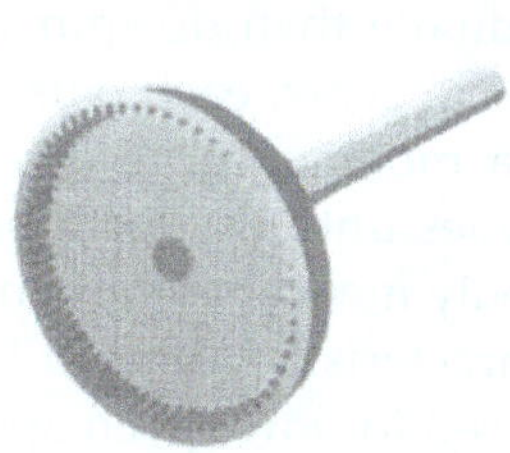

Fig. 4.20

### Internal Gear

An *external gear* is one with the teeth formed on the outer surface of a cylinder or cone. Conversely, an *internal gear* is one with the teeth formed on the inner surface of a cylinder or cone. For bevel gears, an internal gear is one with the pitch angle exceeding 90 degrees. Internal gears do not cause direction reversal.

### Spur

**Fig. 4.21**

### Spur Gear

*Spur gears* or *straight-cut gears* are the simplest type of gear. They consist of a cylinder or disk with the teeth projecting radially, and although they are not straight-sided in form, the edge of each tooth is straight and aligned parallel to the axis of rotation. These gears can be meshed together correctly only if they are fitted to parallel axles.

**Fig. 4.22**

### Helical Gears

*Helical gears* offer a refinement over spur gears. The leading edges of the teeth are not parallel to the axis of rotation, but are set at an angle. Since the gear is curved, this angling causes the tooth shape to be a segment of a helix. Helical gears can be meshed in a *parallel* or *crossed* orientations.

The angled teeth engage more gradually than do spur gear teeth causing them to run more smoothly and quietly. With parallel helical gears, each pair of teeth first make contact at a single point at one side of the gear wheel; a moving curve of contact then grows gradually across the tooth face to a maximum then recedes until the teeth break contact at a single point on the opposite side. In spur gears teeth suddenly meet at a line contact across their entire width causing stress and noise. Spur gears make a characteristic whine at high speeds and cannot take as much torque as helical gears. Spur gears are used for low speed applications and those situations where noise control is not a problem, the use of helical gears is indicated when the application involves high speeds, large power transmission, or where noise abatement is important.

A disadvantage of helical gears is a resultant thrust along the axis of the gear, which needs to be accommodated by appropriate thrust bearings, and a greater degree of sliding friction between the meshing teeth, often addressed with additives in the lubricant.

The gears must have the same pressure angle and normal pitch, however the helix angle and handedness can be different. The relationship between the two shafts is actually defined by the helix angle(s) of the two shafts and the handedness, as defined:

$$E = \beta_1 + \beta_2 \text{ for gears of the same handedness}$$
$$E = \beta_1 - \beta_2 \text{ for gears of opposite handedness}$$

Where $\beta$ is the helix angle for the gear. The crossed configuration is less mechanically sound because there is only a point contact between the gears, whereas in the parallel configuration there is a line contact.

Quite commonly helical gears are used with the helix angle of one having the negative of the helix angle of the other; such a pair might also be referred to as having a right-handed helix.

**Double Helical**

Fig. 4.23

Double helical gears, or *herringbone gear*, overcome the problem of axial thrust presented by "single" helical gears by having two sets of teeth that are set in a *V* shape. Each gear in a double helical gear can be thought of as two standard mirror image helical gears stacked. This cancels out the thrust since each half of the gear thrusts in the opposite direction. Double helical gears are more difficult to manufacture due to their more complicated shape.

The gears become misaligned in the axial direction, the unstable arrangement generates a net force for disassembly of the gear train, while the stable arrangement generates a net corrective force. If the direction of rotation is reversed, the direction of the axial thrusts is reversed, a stable configuration becomes unstable, and *vice versa*.

Stable double helical gears can be directly interchanged with spur gears without any need for different bearings.

**Bevel**

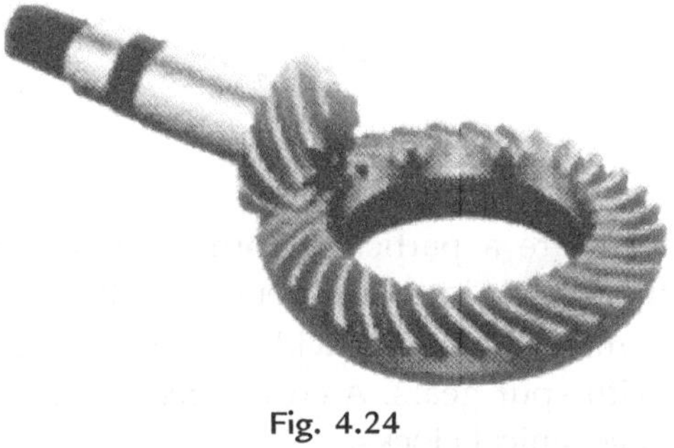

Fig. 4.24

## Bevel Gear

A bevel gear is shaped like a right circular cone with most of its tip cut-off. When two bevel gears mesh their imaginary vertices must occupy the same point. Their shaft axes also intersect at this point, forming an arbitrary non-straight angle between the shafts. The angle between the shafts can be anything except zero or 180°. Bevel gears with equal numbers of teeth and shaft axes at 90° are called *miter* gears.

The teeth of a bevel gear may be straight-cut as with spur gears, or they may be cut in a variety of other shapes. *Spiral bevel gear* teeth are curved along the tooth's length and set at an angle.

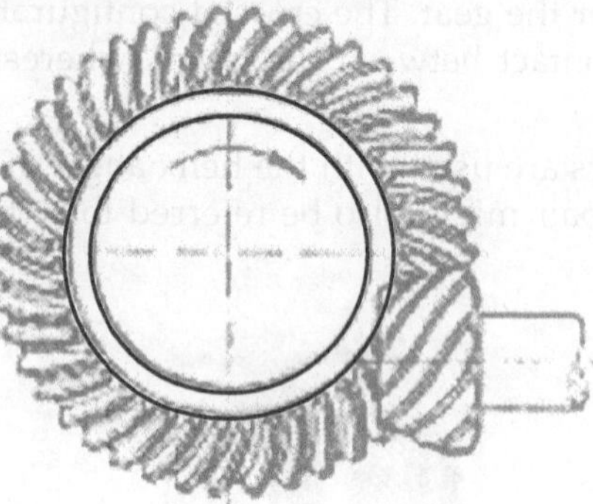

Fig. 4.25

## Spiral Bevel Gear

Hypoid gears resemble spiral bevel gears except the shaft axes do not intersect. The pitch surfaces appear conical but, to compensate for the offset shaft, are in fact hyperboloids of revolution. Hypoid gears are almost always designed to operate with shafts at 90°. The shaft is offset to, relative to the angling of the teeth, contact between hypoid gear teeth may be even smoother and more gradual than with spiral bevel gear teeth. Also, the pinion can be designed with fewer teeth than a spiral bevel pinion, with the result that gear ratios of 60 : 1 and higher are feasible using a single set of hypoid gears. This style of gear is most commonly found in mechanical differentials.

## Crown Gear

Fig. 4.26

*Crown gears or contrate* gears are a particular form of bevel gear whose teeth project at right angles to the plane of the wheel; in their orientation the teeth resemble the points on a crown. A crown gear can only mesh accurately with another bevel gear, although crown gears are sometimes seen meshing with spur gears. A crown gear is also sometimes meshed with an escapement such as found in mechanical clocks.

### Worm Drive

*Worm gears resemble screws.* A worm gear is usually meshed with an ordinary looking, disk-shaped gear, which is called the *gear*, *wheel*, or *worm wheel*.

Worm-and-gear sets a simple and compact way to achieve a high torque, low speed gear ratio. For example, helical gears are normally limited to gear ratios of less than 10 : 1 while worm-and-gear sets vary from 10 : 1 to 500 : 1. A disadvantage is the potential for considerable sliding action, leading to low efficiency.

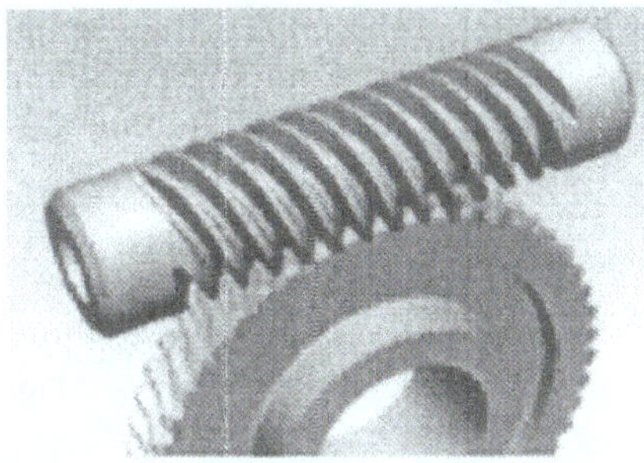

Fig. 4.27

### Gear Profile

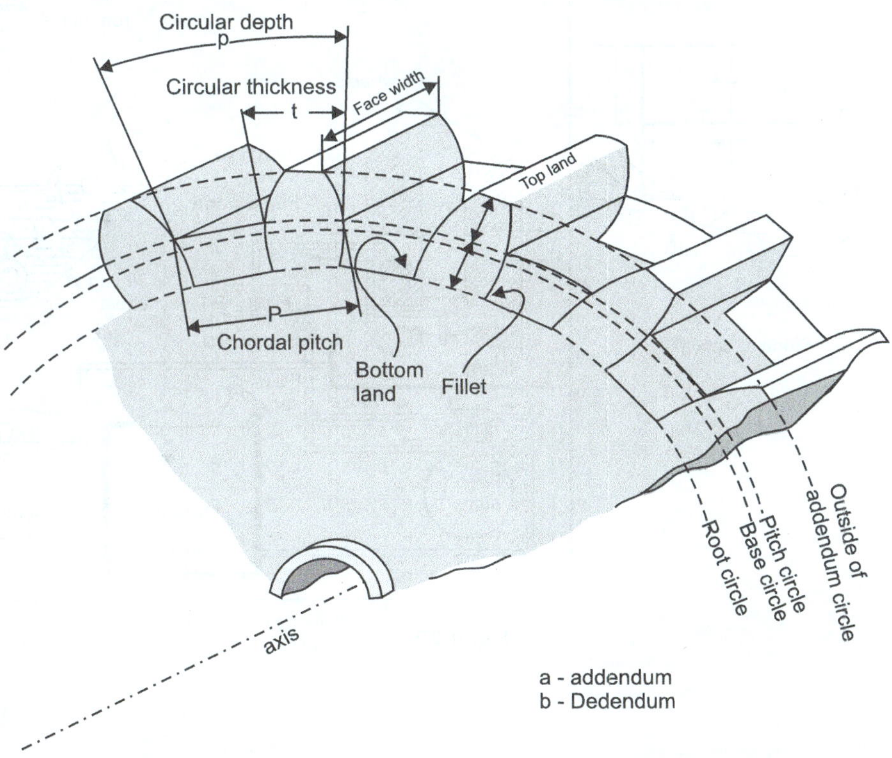

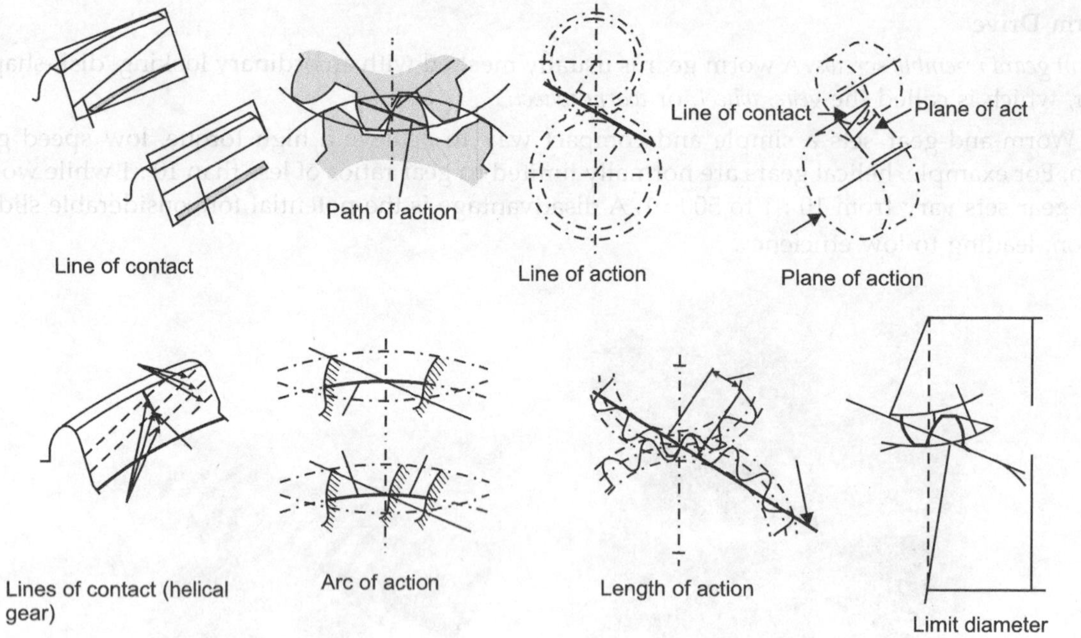


Fig. 4.28

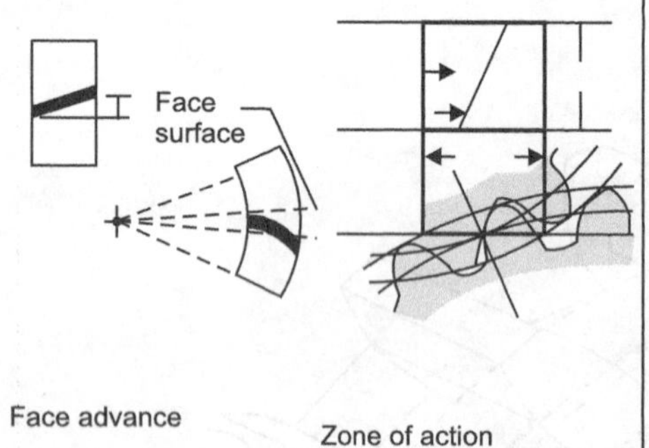


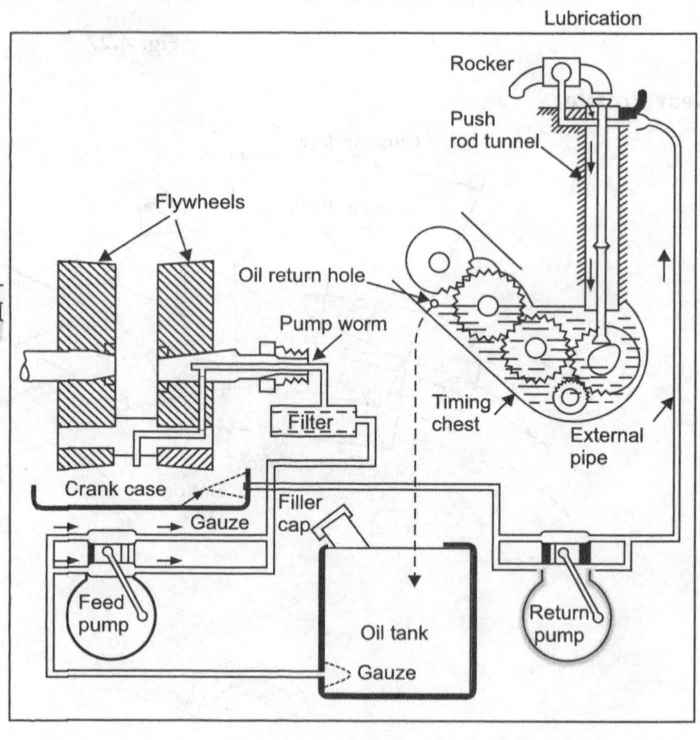


Fig. 4.29

### Cam

The cam as a device that translates from circular to reciprocating motion. Of an automobile, which takes the rotary motion of the engine and translates it into the reciprocating motion necessary to operate the intake and exhaust valves of the cylinders.

The opposite operation, translation of reciprocating motion to circular motion, is done by a crank. An example is the crankshaft of a car, which takes the reciprocating motion of the pistons and translates it into the rotary motion necessary to operate the wheels.

Cams can also be viewed as information-storing and transmitting devices. Examples are the cam-drums that direct the notes of a music box or the movements of a screw machine's various tools and chucks.

### Displacement Diagram

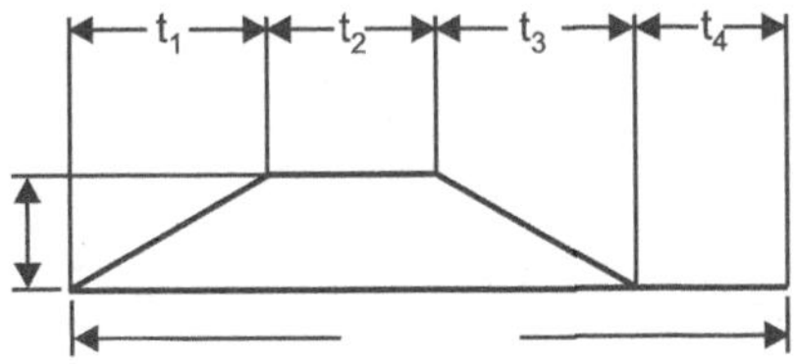


Fig. 4.30

Certain cams can be characterized by their displacement diagrams, which reflect the changing position a roller follower (a shaft with a rotating wheel at the end) would make as the cam rotates about an axis. These diagrams relate angular position, usually in degrees, to the radial displacement experienced at that position. Displacement diagrams are traditionally presented as graphs with non-negative values. A simple displacement diagram illustrates the follower motion at a constant velocity rise followed by a similar return with a dwell in between. The rise is the motion of the follower away from the cam center, dwell is the motion where the follower is at rest, and return is the motion of the follower toward the cam center.

### Plate Cam

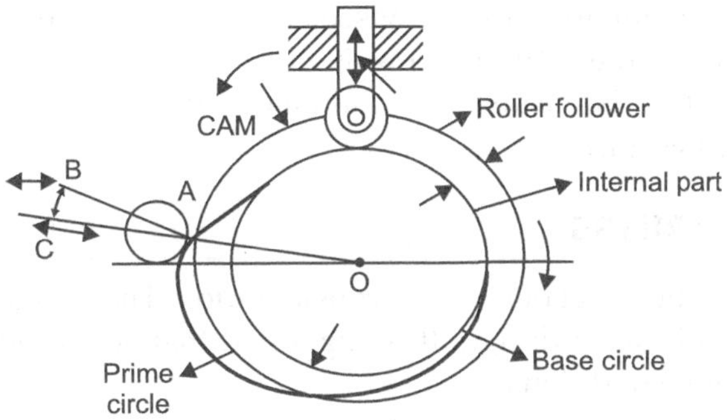


Fig. 4.31

### Cam Profile

The most commonly used cam is the plate cam which is cut out of a piece of flat metal or plate. Here, the follower moves in a plane perpendicular to the axis of rotation of the camshaft. Several key terms are relevant in such a construction of plate cams: base circle, prime circle (with radius equal to the sum of the follower radius and the base circle radius), pitch curve which is the radial curve traced out by applying the radial displacements away from the prime circle across all angles, and the lobe separation angle (**LSA**), the angle between two adjacent intake and exhaust cam lobes.

Chapter 5

# BEARINGS

## 5.1 INTRODUCTION

Bearings are machine elements whose function is to bear load and promote smooth relative motion at low friction between solid surfaces. The surfaces might be in direct contact, but if they are not, the lubricant film separating them can be liquid, gaseous, or solid. Bearing which are used for the movements of the surfaces. The maximum film pressure is of the same order of magnitude as the *specific bearing load*, defined by $P = W/A$, where $W$ is the external load and A is the projected (normal to the load) bearing area. The elastic deformation of the solid surfaces forms an essential component of the analysis of these bearings. There are two principal ways of creating and maintaining a load-carrying film between solid surfaces in relative motion when the film is generated and maintained by the viscous drag of the surfaces themselves, as they are sliding relative to one another. The bearing is *externally pressurized*, and it operates in the *hydrostatic mode*, when the film is created and maintained by an external pump that forces the lubricant between the solid surfaces.

## 5.2 DESIGN OF BEARINGS

The design of Bearing basis the effect of material consideration. The design data below is provided as an introduction to nylon bearing design. It is suggested that actual field testing be conducted under proper environmental conditions.

### PV Value

The frictional build-up of heat is a major consideration in the design of nylon bearings. The two significant factors that affect heat generation are unit pressure ($P$) and surface velocity ($V$). Pressure velocity ($PV$), therefore, is the product of unit pressure and surface velocity.

Surface velocity for sleeve-type bearings can be computed using the following:

$$V = .262 \times \text{rpm} \times D$$

rpm represents shaft revolutions per minute, and $D$ is the shaft diameter in inches.

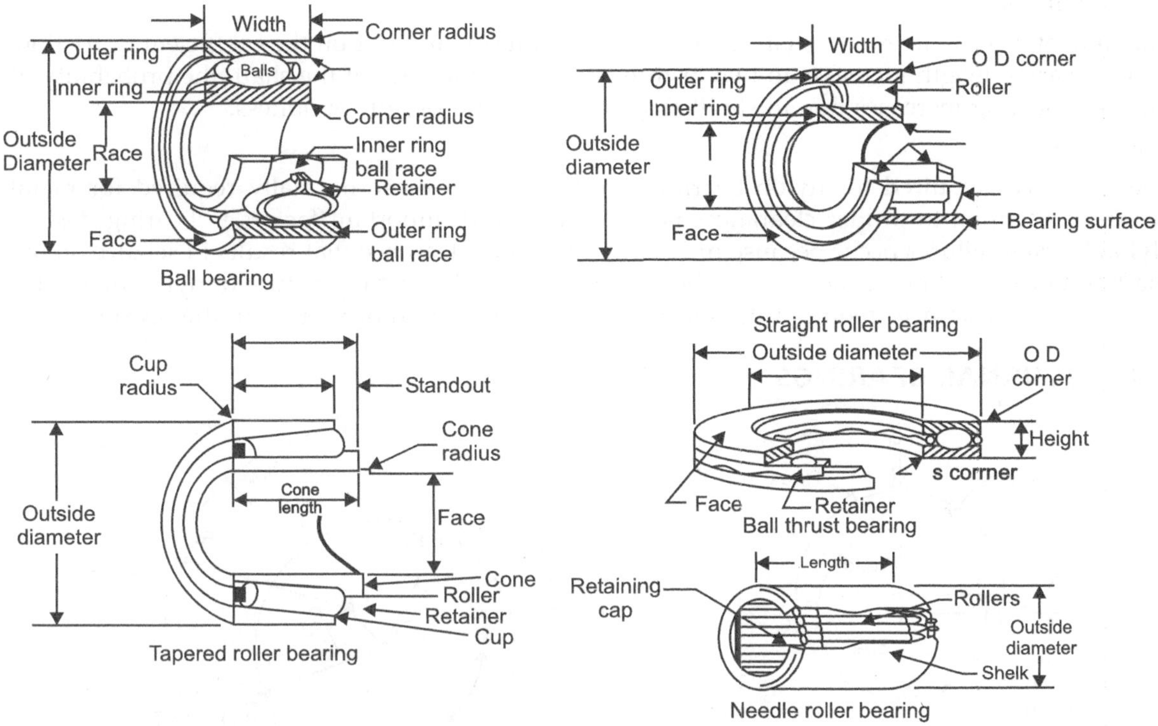


**Fig. 5.1:** Rolling bearing features and terminology

Pressure can be computed for flat bearing units by dividing the total load in pounds by contact area expressed in square inches:

$$P = \text{Total Load/Contact Area}$$

The total load is divided by the projected area of the bearing surface (bearing inner diameter in inches × bearing length in inches). Once pressure and surface velocity are known, *PV* can be determined by using the following:

$$PV = P \times V$$

## Ambient Temperature Correction

When the ambient temperature is higher or lower than 28°C, the PV capabilities change. The following graph indicates the relationship between ambient temperatures and the necessary PV value of bearing.

## Machining Tolerance

Proper running clearances for bearing inner diameter (I.D) and outer diameter (O.D) can be assured by holding tight machining tolerances. It is recommended that inner diameter be machined to plus best commercial tolerance, minus 0.00 inches. O.D should be machined to plus or minus best commercial tolerance.

## Nominal Wall Thickness

For bearings that will be subjected to severe impact, maximum walls are suggested, while minimum walls can be used for bearings that operate near the material's maximum recommended PV.

Nominal wall thickness is the difference between shaft diameter and the housing inner diameter divided by 2.

### Bearing Shape

The ratio of bearing length to shaft diameter has an important effect on the coefficient of friction. As the bearing length increases to two or three times the diameter of the shaft, the probability of localized heating increases due to slight shaft vibration and out-of-roundness.

### Clearance

The clearances required for nylon bearings are larger than those normally required for metal bearings, attention to proper clearances is one of the most important factors in bearing design. Nylon bearing failures occur because of insufficient initial clearance that results in seizure on the shaft and even melting of the bearing. Because of the self-dampening nature of nylon materials, the increased clearances required do not result in shaft vibration or scoring of the bearing.

## 5.3 JOURNAL BEARINGS

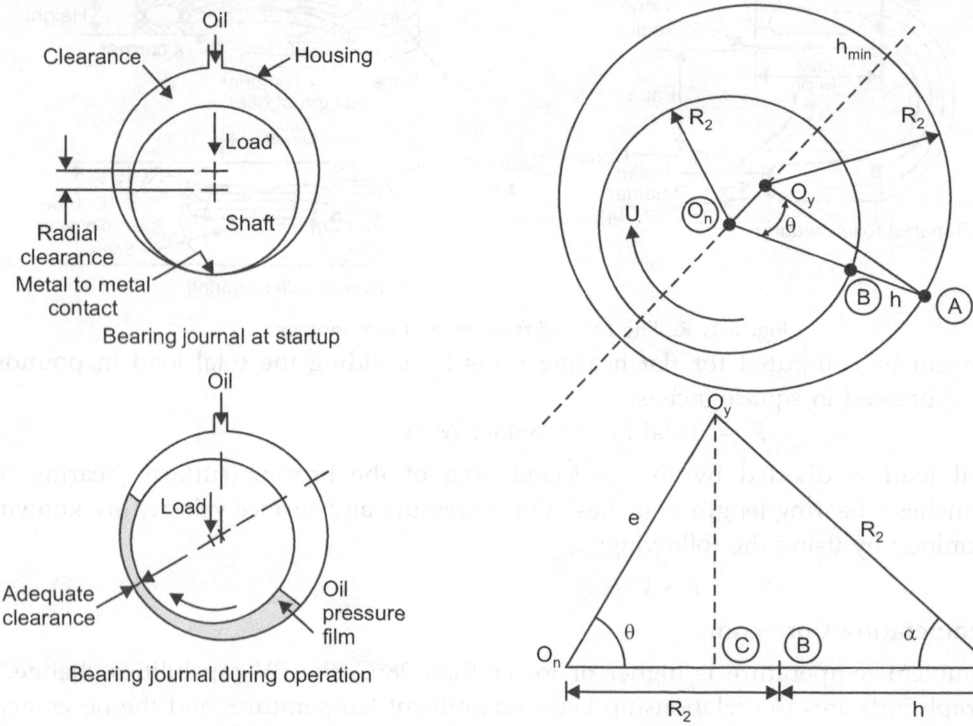


Fig. 5.2

Journal bearings are very common engineering components and are used in almost all types of machinery. Combustion engines and turbines virtually depend on journal bearings to obtain high efficiency and reliability. A journal bearing consists of a shaft rotating within a stationary bush. The hydrodynamic film which supports the load is generated between the moving surfaces of the shaft and the bush.

### Bearing Geometry

Bearing geometry is defined in the first step of the analysis. That the angle θ is very small. From inspection of the triangle **O, SOB, A** can be written:

$$O_SA = O_SC + CA = O_SB + BA$$

or $$O_SA = e\cos\theta + R_1\cos\alpha = R_2 + h$$

thus $$h = e\cos\theta + R_1\cos\alpha - R_2 \text{ Applying sine rules}$$

$$\frac{e}{\sin\alpha} = \frac{R_2}{\sin\theta} \quad \text{and} \quad \sin\alpha = \frac{e}{R_1}\sin\theta$$

$$\sin^2\alpha + \cos^2\alpha = 1 \quad \text{say} \quad \delta\theta\cos\alpha = \sqrt{1-\sin^2\alpha} \quad \text{or} \quad \sqrt{1-\left(\frac{e}{R_1}\right)^2\sin^2\theta}$$

So $$\frac{e}{R_1} \cong 1 \quad \text{then} \quad h = e\cos\theta + R_1 - R_2 = e\cos\theta + C \text{ (C) clearance}$$

or $$h = C(1+\varepsilon\cos\theta)$$

**Pressure Distribution**

In journal bearings where the axial length is less than the shaft diameter, the pressure gradient along the '$y$' axis is much larger than the pressure gradient along the '$x$' axis (circumferential direction). The narrow bearing approximation can therefore be used. This approximation gives accurate results for **L/D < 1/3**. The one-dimensional Reynolds equation for the narrow bearing:

$$P = \frac{3U_n}{h^3}\frac{dh}{dn}\left(y^2 - \frac{L^2}{4}\right)$$

$L$ = length of Bearing

$x$ = Angular displacement

$x$ = $R\theta$ and $dx = Rd\theta$ when differenciate

$$\therefore \quad P = \frac{3U_n}{h^3}\frac{dh}{d\theta}\left(y^2 - \frac{L^2}{4}\right) \text{ Diff. } dh = -c\varepsilon\sin\theta\, d\theta$$

$$\therefore \quad \frac{dh}{d\theta} = -c\varepsilon\sin\theta \text{ Pressure distribution}$$

$$(P) = \frac{3U_n\,\varepsilon\sin\theta}{RC^2(1+\varepsilon\cos\theta)^3}\left(\frac{L^2}{4} - y^2\right)$$

**Load Capacity**

The total load that the bearing will support is found by integrating the pressure around the bearing. Load is usually calculated from two components, one acting along the line of shaft and bush centres and a second component perpendicular to the first. The angle is known as the **'attitude angle'** and results in the position of minimum film thickness lying some distance from where the load-line intersects the shaft and bush. In cylindrical form, the true shape of the hydrodynamic film is represented when unwrapped or flat. In this way the form most convenient for film visualization is obtained. The increment of force exerted by the hydrodynamic pressure on the element of area is:

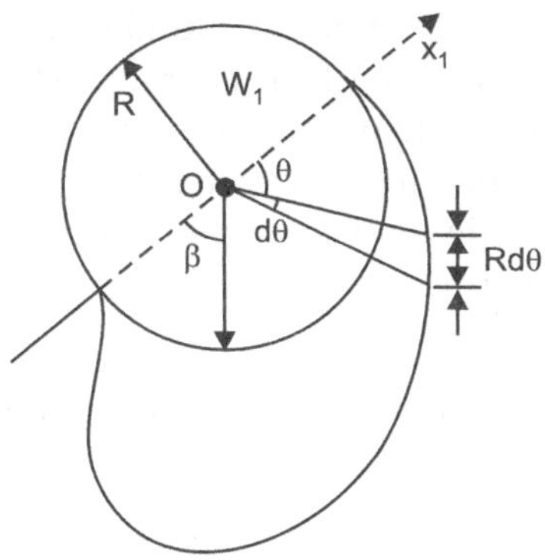

**Fig. 5.3:** Pressure profile

The total component $w_x$ in direction of line of centre is integral

$$w_x = \int_0^{\pi}\int_{-\frac{L}{2}}^{+\frac{L}{2}} pR\,d\theta\,dy\cos\theta$$

$$w_y = \int_0^{\pi} \int_{-\frac{L}{2}}^{+\frac{L}{2}} PR\, d\theta\, dy \sin\theta$$

living out the limits

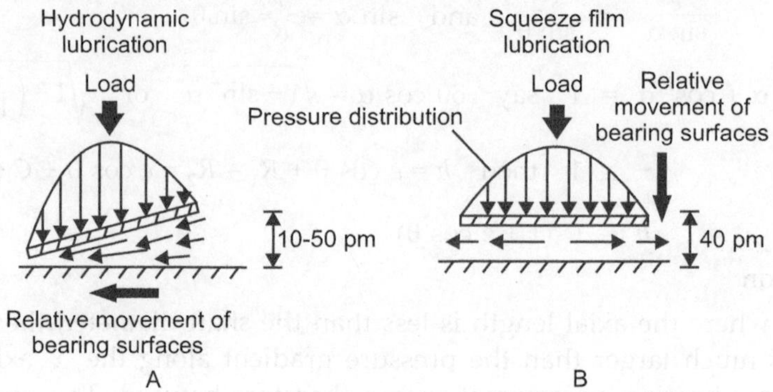


**Fig. 5.4**

$$Aw_y = \iint P \sin\theta\, R\, d\theta\, dy = \iint \frac{3U\, \eta e \sin^2 R\, d\theta}{RC^2 (1 + \varepsilon \cos\theta)^3} \left(\frac{L^3}{4} - y^2\right) dy \quad \ldots(1)$$

Cancelling $R$'s top then

$$Aw_x = \iint P \cos\theta\, R\, d\theta\, dy = \iint \frac{3U_\eta\, e \sin\theta \cos\theta\, d\theta}{C^2 (1 + \varepsilon \cos\theta)^3} \left(\frac{L^2}{4} - y^2\right) dy \quad \ldots(2)$$

When integrates:

$$\int_{-\frac{L}{2}}^{+\frac{L}{2}} \left(\frac{L^2}{4} - y^2\right) dy = 2\int_0^{\frac{L}{2}} \left(\frac{L^2}{4} - y^2\right) dy = 2\left|\frac{L^2 y}{4} - \frac{y^3}{3}\right|_0^{\frac{L}{2}} = +\frac{L^3}{6}$$

So
$$W = \frac{U_\eta L^3}{C^2} \left(\frac{\pi^2 \varepsilon^2}{16 (1 - \varepsilon^2)^3} + \frac{\varepsilon^4}{(1 - \varepsilon^2)^4}\right)^{1/2}$$

**Sommerfeld Number**

The Sommerfeld Number is a very important parameter in bearing design since it expresses the bearing load characteristic as a function of eccentricity ratio. Sommerfeld number 'θ' versus eccentricity ratio 'θ'. The curves were computed using the Reynolds boundary condition which is the more accurate. Removal of the bearing shell at positions where hydrodynamic pressure is negligible is a convenient means of reducing friction and the bearings are known as partial arc bearings. The effect on load capacity is negligible except at extremely small eccentricity ratios.

$$E_{\text{optimum}} = 0.7$$

when modifies this parameters becomes

$$U = 2\,\pi RN$$

$$\Delta = \frac{W}{L_\eta 2\pi RN} \left(\frac{C}{R}\right)^2 \quad \text{Introduce } (P) = \frac{W}{2LR}$$

$$\Delta = \frac{P}{N_\eta \pi} \left(\frac{C}{R}\right)^2 \quad \text{Thus} \quad S = \Delta\pi = \frac{P}{N_\eta} \left(\frac{C}{R}\right)^2$$

*Attitude angle*: The value of minimum film thickness can be determined from a knowledge of ε which can be found from the external parameter contained angle between $W$ and line of centre

$$\tan \psi = -\frac{W_y}{W_x}$$

$$\therefore \quad \tan \psi = \frac{\pi}{4}\left(\frac{\varepsilon}{(1-e^2)^{3/2}}\right) \cdot \frac{(1-\varepsilon^2)^3}{\varepsilon} = \frac{\pi}{4}\left[\frac{1-\varepsilon^2}{\varepsilon}\right]^{1/2}$$

$$\tan \psi = \frac{\sin \psi}{\cos \psi} = \frac{(1-\cos^2 \psi)^{1/2}}{\cos \psi} = \frac{(1-\varepsilon^2)^{1/2}}{\varepsilon}$$

which means that if ε and ψ are plotted in polar coordinates the locus of ε would have been semicircle.

Journal bearings are commonly incorporated as integral parts of various machinery with a wide range of design requirements problems associated with practical implementation and operation of journal bearings. For example, in many practical applications the lubricant is fed under pressure into the bearing or there are some critical resonant shaft speeds to be avoided.

## Cavitation

Negative pressures in the hydrodynamic film are predicted when surfaces move apart or mutually sliding surfaces move in a divergent direction. For gases, a negative pressure does not exist and for most liquids a phenomenon known as cavitation occurs when the pressure falls below atmospheric pressure. When the pressure becomes subatmospheric, bubbles of previously dissolved air nucleate on pits, cracks and other surface irregularities on the sliding surfaces and also on dirt particles. If there is a significant drop in pressure, the operating temperature can be sufficient for the lubricant to evaporate. The lubricant vapour accumulates in the bubbles and their sudden collapse is the cause of most cavitation damage. The critical difference between 'gaseous cavitation', *i.e.*, cavitation involving bubbles of dissolved air, and 'vaprous cavitation' is that with the latter, sudden bubble collapse is possible. When a bubble collapses against a solid surface very high stresses, reaching 0.5 [GPa] in some cases, are generated and this will usually cause wear.

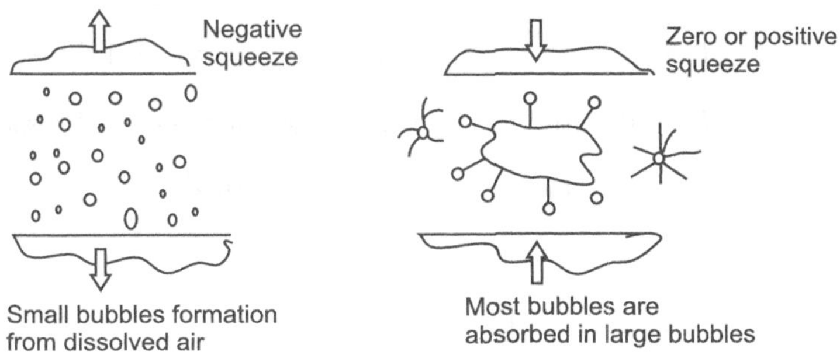


**Fig. 5.5**

Cavitation occurs in liquid lubricated journal bearings, in elastohydrodynamics and in applications other than bearings such as propeller blades. In journal bearings, cavitation causes a series of 'streamers' to form in the film space.

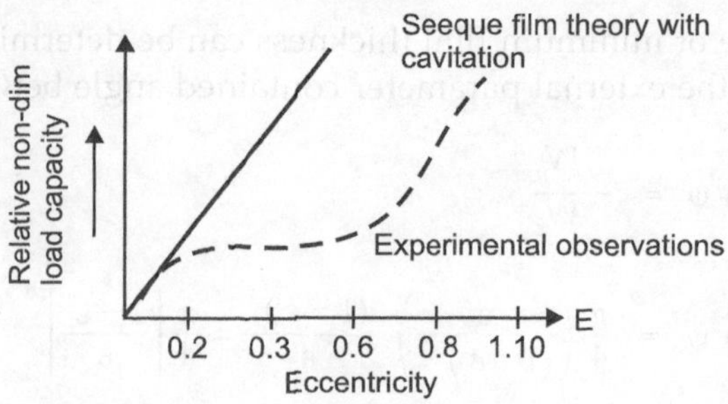


Fig. 5.6

## 5.4 REYNOLDS EQUATION FOR SQUEEZE FILMS

Squeeze film is a term denoting a hydrodynamic film that sustains a negative, *i.e.*, when the opposing surfaces are being squeezed together. An extremely useful characteristic of squeeze films is that they provide increased load capacity when a bearing is suddenly subjected to an abnormally high load. The Reynold's equation with squeeze terms:

$$\frac{\partial}{\partial x}\left(\frac{h^3}{\eta}\frac{\partial P}{\partial x}\right)+\frac{\partial}{\partial y}\left(\frac{h^3}{\eta}\frac{\partial P}{\partial y}\right) = 6\frac{\partial h}{\partial x}+12\frac{\partial h}{\partial t}$$

$$\frac{\partial}{\partial x}\left(h^3\frac{\partial P}{\partial x}\right)+\frac{\partial}{\partial y}\left(h^3\frac{\partial P}{\partial y}\right) = 12\eta\frac{\partial h}{\partial t}$$ When two parallel and infinite long plates

$h \neq f(x)$ and $\frac{\partial P}{\partial y} = 0$ so $\frac{d^2P}{dx^2} = \frac{12\eta}{h^3}\frac{dh}{dt}$ Integrate one

$\frac{dP}{dx} = \frac{12\eta}{h^3}\frac{dh}{dt}x + C_1$ once again integrates

This equation defines the hydrodynamic pressure field when the wedge effect is absent, *e.g.*, when the load vector rotates as mentioned previously. It can be integrated in terms of a specified bearing geometry to provide load capacity, maximum pressure or any other required bearing characteristic in terms of the 'squeeze time' which means the time required for film thickness to decline to some critical minimum value can also be determined by integrating with respect to time.

$P = \frac{6\eta}{h^3}\frac{dh}{dt}x^2 + C_1x + C_2$ Boundary conditions

$P = 0$ and $x \neq \pm B/2$ edges of plates at Pressure Gradient

$\frac{dP}{dx} = 0$ at $x = 0$ substituting these

$$C_1 = 0$$

$$C_2 = -\frac{6\eta B^2}{4h^3}\frac{dh}{dt}$$

Substituting $C_1$ and $C_2$ gives pressure distribution:

$P = \frac{6\eta}{h^3}\frac{dh}{dt}\left[x^2 - \frac{B^2}{4}\right]$ where the terms $\frac{dh}{dt}$ is negative the pressure distribution over the Bearing for positive seeque directions.

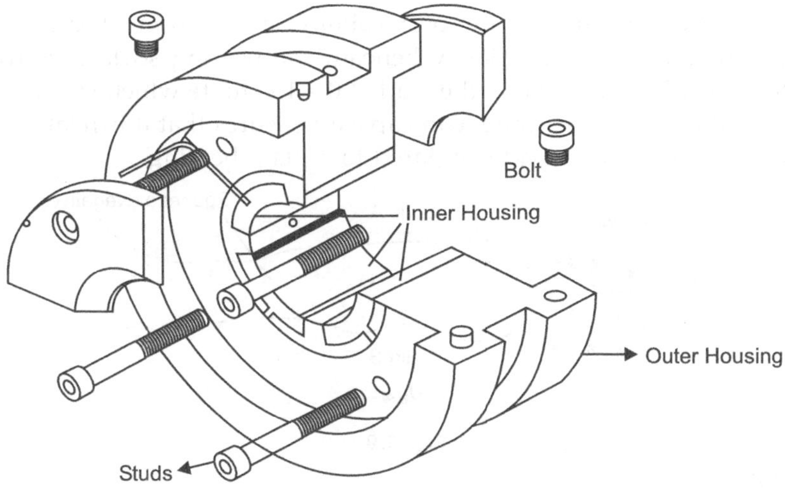


Fig. 5.7

## 5.5 POROUS BEARINGS

Porous bearings are intended to provide hydrodynamic lubrication without the need for an external lubricant supply. These bearings consist of a cylindrical bush made of compacted metal powder which has been impregnated with a lubricant. Transient squeeze between rough surfaces in close contact mineral oil and the metal powder is usually bronze. The mechanism of oil recirculation around the bush during operation. The hydrodynamic lubrication of porous bearings can be solved by incorporating a leakage term **'$K_p$'** into the right-hand side of the standard Reynolds equation

$$\frac{\partial}{\partial x}\left(h^3\frac{\partial P}{\partial x}\right)+\frac{\partial}{\partial y}\left(h^3\frac{\partial P}{\partial y}\right) = \eta\left[6\frac{\partial h}{\partial x}+K_p\left(\frac{\partial P}{\partial z}\right)_0\right]$$

where $K_p$ = Constant depends on permeability of porous Bearing

$\left(\frac{\partial P}{\partial z}\right)_0$ = Pressure gradient in Porous material

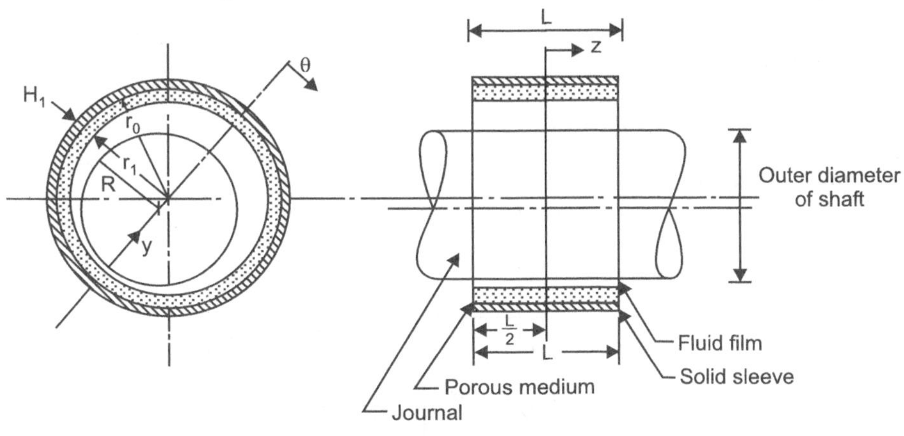


**Fig. 5.8:** Bearing geometry and co-ordinates

$K_p$ is a constant which depends on the permeability of the porous bearing; is the pressure gradient in the porous material immediately beneath the bearing surface. Porous bearings as predicted by hydrodynamic theory is that the load at θ = 1 is finite which is in contrast to infinite load for a plain journal bearing. The finite load capacity ensures that it is relatively easy to cause seizure in a porous bearing by overload compared to a plain bearing.

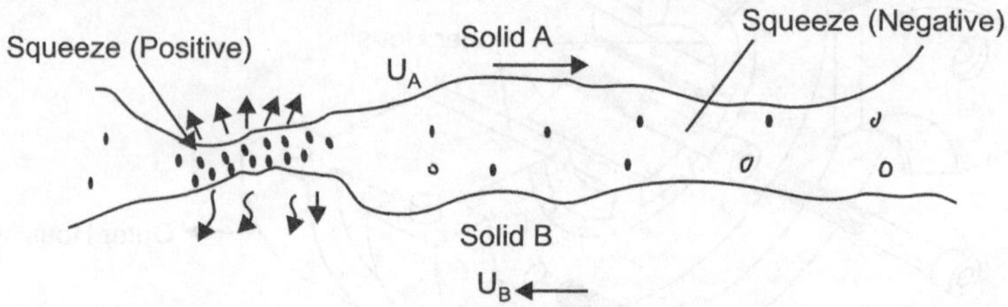


Fig. 5.9

## Reynolds Equation

The oil in a journal bearing always leaked out of a hole located beneath the load. The leakage of oil was a nuisance so the hole was plugged first with a cork, which still allowed oil to ooze out, and then with a hard wooden bung. The hole was originally placed to allow oil to be supplied into the bearing to provide 'lubrication'. All hydrodynamic lubrication can be expressed mathematically in the form of an equation which was originally derived by Reynolds and is commonly known throughout the literature as the 'Reynolds equation'. There are two conditions for the occurrence of hydrodynamic lubrication:

- two surfaces must move relatively to each other with sufficient velocity for a load carrying lubricating film to be generated and the relation between them.
- surfaces must be inclined at some angle to each other, *i.e.*, if the surfaces are parallel a pressure field will not form in the lubricating film to support the required load of the object.

There are two exceptions to this last rule: hydrodynamic pressure can be generated between parallel stepped surfaces or the surfaces can move towards each other.

$$Pdydz + \left(\tau_x + \frac{\partial z_x}{\partial z} dz\right] dx\, dy = \left(P + \frac{\partial P}{\partial x} dx\right) dy\, dz + \tau_x\, dx\, dy$$

which gives the relation:

$$\frac{\partial z_x}{\partial z} dx\, dy\, dz = \frac{\partial P}{\partial x} dx\, dy\, dz$$

Assuming that $dx\, dy\, dz \neq 0$

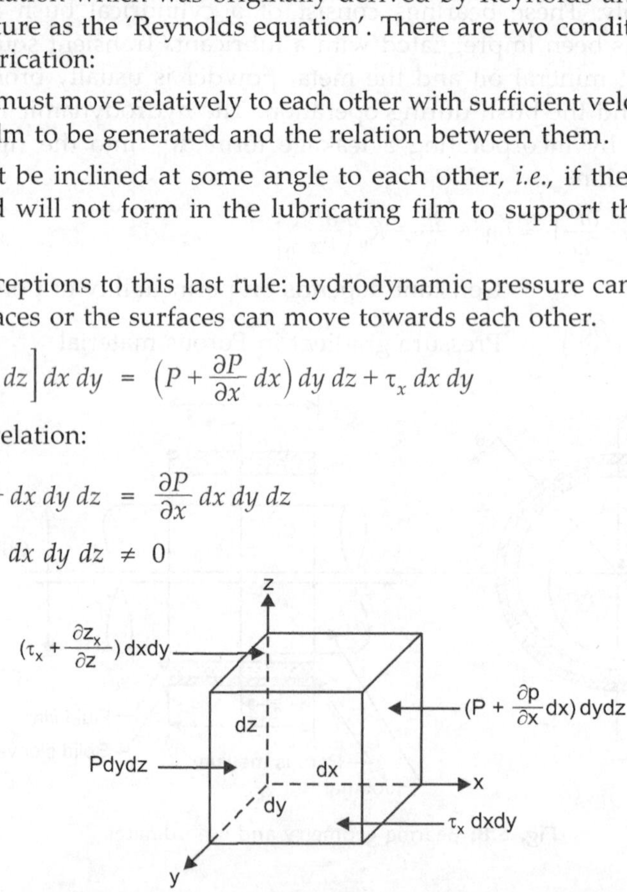


Fig. 5.10

The equilibrium conditions for forces acting '$x$' direction is:

$$\frac{\partial z_x}{\partial z} = \frac{\partial P}{\partial x} \quad \text{for } (x) \text{ direction}$$

when forces acting '$y$' direction:

$$\frac{\partial z_y}{\partial z} = \frac{\partial P}{\partial y}$$

when forces in '$z$' direction:

$$\frac{\partial P}{\partial z} = 0$$ when the shear stresses '$\tau$' can be expressed in terms of dynamic viscosity and shear stress:

$$\tau_x = \eta \frac{u}{h} = \eta \frac{dv}{dz} \qquad \text{...sliding velocity '}x\text{' direction}$$

$$\tau_y = \eta \frac{V}{h} = \eta \frac{\partial v}{\partial z} \qquad \text{...sliding velocity '}y\text{' direction}$$

$$\frac{\partial P}{\partial x}\partial z = \partial\left(\eta \frac{\partial v}{\partial z}\right)$$ and integrates these:

$$\left(\frac{\partial P}{\partial x}z + C_1\right) = \eta \frac{\partial v}{\partial z}$$ again integrates;

$$\frac{\partial P}{\partial x}\frac{z^2}{2} + C_1 z + C_2 = \eta u$$

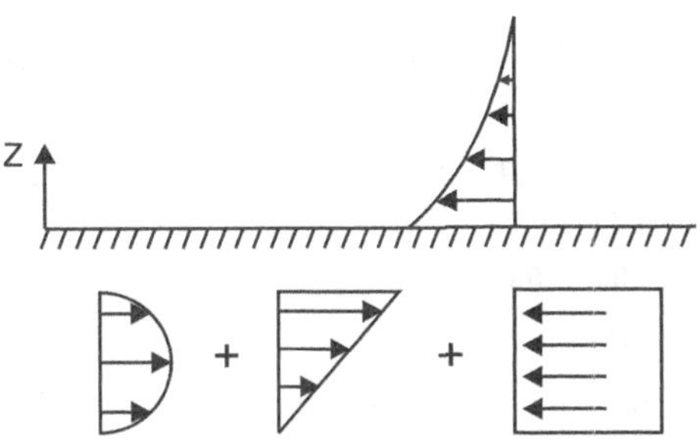


**Fig. 5.11**

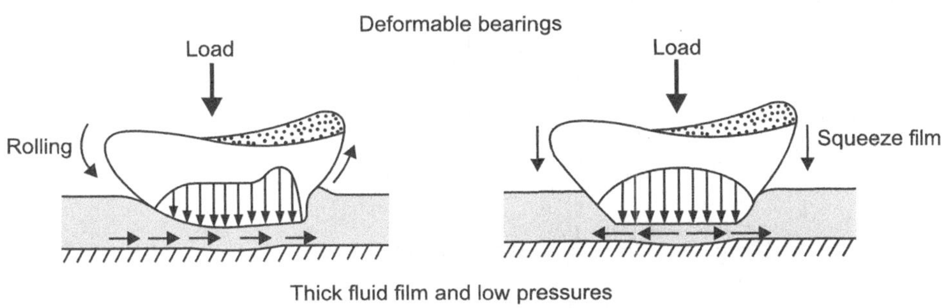


**Fig. 5.12**

**Continuity of Flow in a Column**

Consider a column of lubricant that the lubricant flows into the column horizontally at rates of '$q_x$' and '$q_y$' and out of the column at rates per unit length and width respectively. In the vertical direction the lubricant flows into the column at the rate of '$w_0 dxdy$' and out of the column at the rate of '$w_h dxdy$', where '$w_0$' is the velocity at which the bottom of the column moves up and '$w_h$' is the velocity at which the top of the column moves up. The principle of continuity of flow requires that the influx of a liquid must equal its efflux from a control volume under steady conditions. If the density of the lubricant is constant then the following relation applies:

$$q_z\, dy + q_y\, dx + w_0\, dx\, dy = \left[q_x + \frac{\partial q_x}{\partial x} dx\right) dy + \left(q_y + \frac{\partial q_y}{\partial y} dy\right) dx + w_h\, dx\, dy$$

Simplifying this relation:

$$\frac{\partial q_x}{\partial x} dx\, dy + \frac{\partial q_y}{\partial y} dx\, dy + (w_h - w_0)\, dx\, dy = 0$$

Since $dxdy \neq 0$ then

$$\frac{\partial q_x}{\partial x} + \frac{\partial q_y}{\partial y} + (w_h - w_0) = 0$$ flow rates per unit length $q_x$ and $q_y$ then

$$q_x = \int_0^h \mu\, dz \quad \text{and} \quad q_y = \int_0^h v\, dz$$

Substituting for '$u$' from equation:

$$\therefore \quad q_x = \left| \frac{z^3}{3} - \frac{z^2 h}{2} \right| \frac{\partial P}{2\eta\, \partial x} + (v_1 - v_2)\frac{z^2}{2h} + u_2 z \Big|_0^h$$

$$\therefore \quad q_x = -\frac{h^3}{12\eta}\frac{\partial P}{\partial x} + (v_1 + v_2)\frac{h}{2} \quad \text{In } x \text{ direction}$$

$$\therefore \quad q_y = -\frac{h^3}{12\,\eta}\frac{\partial P}{\partial y} + (v_1 + v_2)\frac{h}{2} \quad \text{In } y \text{ direction}$$

$$\frac{\partial}{\partial x}\left[-\frac{h^3}{12\eta}\frac{\partial P}{\partial x} + (u_1 + u_2)\frac{h}{2}\right] + \frac{\partial}{\partial y}\left[-\frac{h^3}{12\eta}\frac{\partial P}{\partial y} + (v_1 + v_2)\frac{h}{2}\right] + (w_h - w_0) = 0$$

Defining

$$U = u_1 + u_2$$

$$V = v_1 + v_2$$

$$\text{so} \; -\frac{\partial}{\partial x}\left[\frac{h^3}{12\eta}\frac{\partial P}{\partial x}\right] + \frac{U}{2}\frac{dh}{dx} - \frac{\partial}{\partial y}\left[\frac{h^3}{12\eta}\frac{\partial P}{\partial y}\right] + \frac{V}{2}\frac{dh}{dx} + (w_h - w_0) = 0$$

Further rearranging and simplifies:

$$\frac{\partial}{\partial x}\left(\frac{h^3}{\eta}\frac{\partial P}{\partial x}\right] + \frac{\partial}{\partial y}\left(\frac{h^3}{\eta}\frac{\partial P}{\partial y}\right] = 6\left[U\frac{dh}{dx} + V\frac{dh}{dy}\right] + 12\,(w_h - w_0)$$

Pressure distribution:

$$\frac{\partial}{\partial y}\left(h^3\frac{\partial P}{\partial y}\right) = 6U\eta\frac{dh}{dx} \quad \text{Since } h \neq f(y)$$

$$\frac{d^2P}{dy^2} = \frac{6U\eta}{h^3}\frac{dh}{dx} \quad \text{Integrating these}$$

$$\frac{dP}{dy} = \frac{6U\eta}{h^3}\frac{dh}{dx}\, y + C_1 \quad \text{and again gives:}$$

$$P = \frac{6U\eta}{h^3}\frac{dh}{dx}\frac{y^2}{2} + C_1 y + C_2 \quad \text{Boundary conditions}$$

$$P = 0 \quad \text{at } y = \pm L/2 \quad \text{or} \quad \frac{dP}{dy} = 0 \text{ at } y = 0$$

$$C_1 = 0 \text{ when constant pressure gradiant}$$

$$C_2 = -\frac{3U\eta}{h^3}\frac{dh}{dx}\frac{L^2}{4}$$

$$\text{So pressure distribution is} = (P) = \frac{3U\eta}{h^3}\frac{dh}{dx}\left(y^2 - \frac{L^2}{4}\right)$$

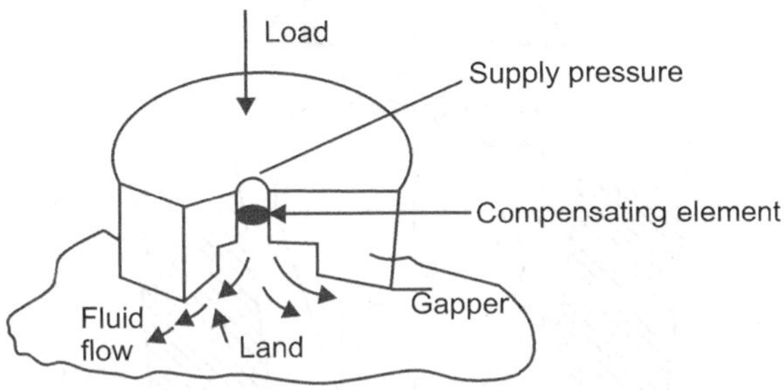


Fig. 5.13

## 5.6 HYDROSTATIC BEARINGS

Hydrostatic bearings have a wide range of characteristics and need to be carefully controlled for optimum effect. If it is possible to generate films similar to hydrodynamic films.

### Hydrostatic Bearing Analysis

The analysis of hydrostatic bearings is much simpler than the analysis of hydrodynamic bearings. It is greatly simplified by the condition that the surfaces of these bearings are parallel.

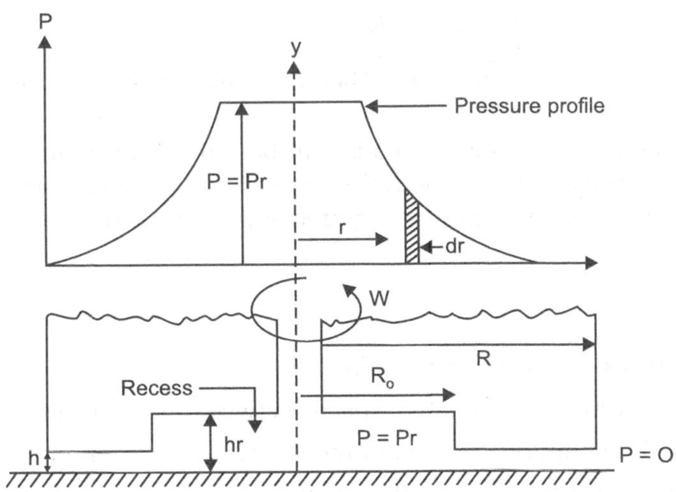


Fig. 5.14

**Flat Circular Hydrostatic Pad Bearings**

Consider, as an example, a flat circular hydrostatic pad bearing with a central recess as shown:

$$q_x = \frac{-h^3}{12\eta}\frac{\partial P}{\partial x}$$

$$Q = \frac{-h^3}{12\eta}\frac{\partial P}{\partial r}2\pi r$$

Rearrainging and integrating these equations.

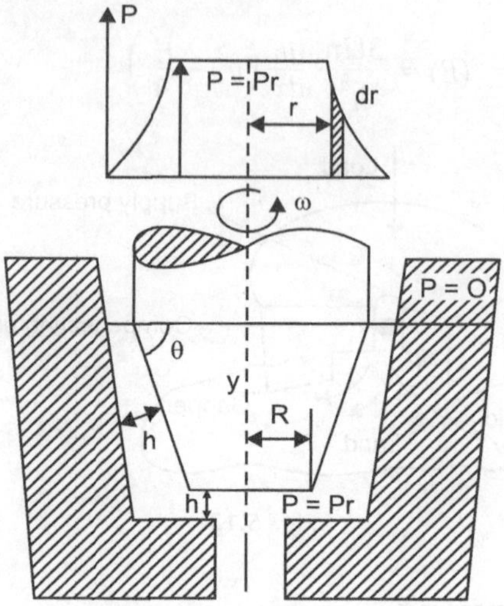


Fig. 5.15

**Pressure Distribution**

The pressure distribution can be calculated by considering the lubricant flow in a bearing. For a bearing supplied with lubricant under pressure, the flow rate given by equation becomes:

$$P = \frac{-6\eta Q}{\pi h^3}\int\frac{dr}{r} = \frac{-6\eta Q}{\pi h^3} l_n r + C \quad \text{where boundary conditions}$$

$$P = 0 \quad \text{at} \quad r = R$$

$$\therefore \quad C = \frac{6\eta Q}{\pi h^3} l_n R \quad \text{Then pressure distribution for Hydrostatic bearing}$$

The net effect of the action of the two pads will be a restoring moment on the runner. An externally pressurized bearing equipped with flow *restrictors* is called a *compensated* bearing. Flow restrictor design influences bearing stiffness, required supply pressure, required pumping power, and lubricant flow.

$$\therefore \quad P = \frac{6\eta Q}{\pi h^3} l_n\left(\frac{R}{r}\right)$$

Among the advantages of hydrostatic bearings are:

- Low friction (vanishing with relative speed) between the surfaces
- Unaffected by discontinuous motion of the runner
- Exact positioning of the runner, as film thickness can be controlled accurately

- Capable of supporting heavy loads
- Only moderate temperature rise across pad because of large film thickness
- Continuous outflow of lubricant prevents ingress of dirt.

## 5.7 GAS LUBRICATED BEARINGS

### Aerostatic Bearings

The mechanism of film generation is the same as in liquid bearings. Gas lubricated bearings offer some advantages such as:

- gas viscosity increases with temperature thus reducing heating effects during overload or abnormal operating conditions,
- some gases are chemically stable over a wider temperature range than hydrocarbon lubricants,
- a non-combustible gas eliminates the fire hazard associated with hydrocarbons,
- if air is selected as the hydrostatic lubricant, then it is not necessary to purchase or recycle the lubricant,
- gases can offer greater cleanliness and non-toxicity than fluid lubricants.

Aerostatic bearings are used in the precision machine tool industry, metrology, computer peripheral devices and in dental drills, where the air used in the bearing also drives the drill. They are particularly useful for high speed applications and where precision is required since very thin film thicknesses are possible. Load capacity of gas lubricated bearings or aerostatic bearings is much lower than the load capacity of fluid lubricated bearings of the same size.

$$P = P_r\left[1 - \frac{ln\,(r \mid R_0)}{ln\,(r \mid R_0)}\right]\left[1 - \frac{P_0}{P_r}\right]^{0.5}$$

### Pressure Distribution

The pressure distribution is found from a simplified form of the compressible Reynolds equation for radial coordinates with angular symmetry and negligible sliding velocity. The derivation of the pressure distribution can be found in the final form of the equation for pressure at any given radial position between the recess and bearing edge is:

where

$$Q = \frac{\pi h^3}{12\eta\, ln\,(R/R_o)} \frac{P_r^2 - P_s^2}{P_r^2}$$

$h$ = film thickness

$\eta$ = gas dynamic viscosity

$$k = \frac{W}{P_r \pi R^2}$$

$w$ = bearing load capacity

$P_r$ = recess pressure

$R$ = outer radius of bearing

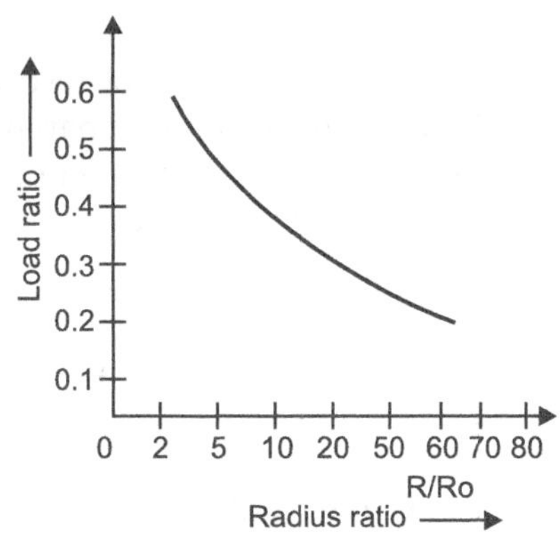


Fig. 5.16

## 5.8 THRUST BEARINGS

**Single Acting**

Analysis of the circular thrust bearing with a central pocket is available from several sources and has been verified in Experiments. Integration of the pressure distribution in this bearing gives the load as a function of pocket pressure:

$$W = \frac{\pi (P_p - P_0)}{2} \left[ \frac{R^2 - R_0^2}{ln\,(R/R_0)} \right]$$

A restrictor in the supply line forces the bearing flow to assume the characteristic of the restrictor. In the flow-vs.-load relations illustrated in since the pocket pressure in these equations is a linear function of load.

$$Q = \frac{\pi h^3 \,[P_p - P_0]}{6\mu \; ln\,(R/R_0)}$$

By equating each restrictor flow equation with the flow equation for this bearing, load capacity expressions can be derived as below.

Orifice restriction:

$$W = \frac{\pi\, k_2\, k_3^2}{4h^6} \left[ -1 + \sqrt{\frac{1 + 4P_S h^6}{k_3^2}} \right]$$

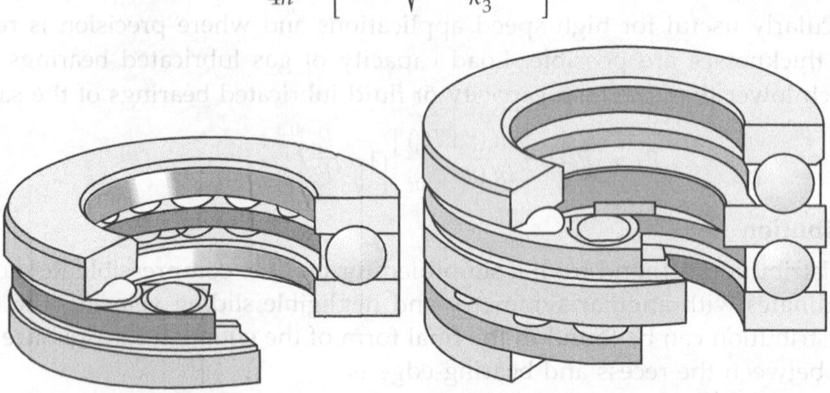

**Fig. 5.17**

When the three types of restrictors are compared at a common operating point, differences in the film stiffnesses are apparent the constant-flow bearing is stiffest, followed by the orifice-restricted and capillary-restricted. The constant flow bearing has a maximum load, set by the maximum supply pressure.

Constant flow:

$$W = 3\mu\, Q_K \left[ \frac{R^2 - R_0^2}{h^3} \right]$$

$$\text{Capillary Restriction} = W = \frac{\pi\,[R^2 - R_0^2]\, P_S}{2\left( \frac{64\, L_c h^3}{3\, d_c^4} + \ln R/R_0 \right)}$$

## Double Acting

Two thrust bearings are often used in opposition. Each bearing is calculated independently over the possible range of film thickness, and the load, flow, and power losses are combined for the assembly.

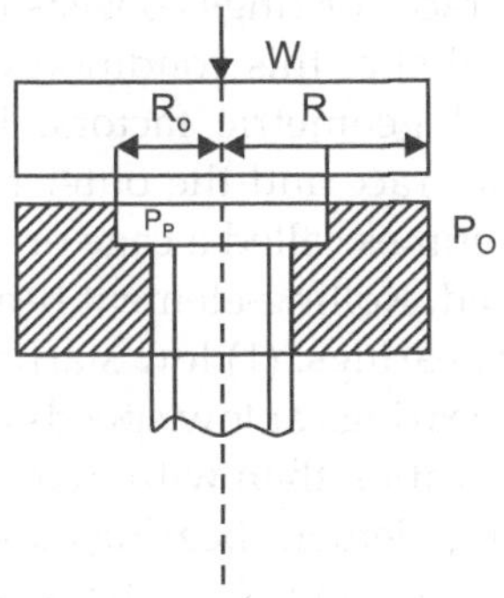


**Fig. 5.18**

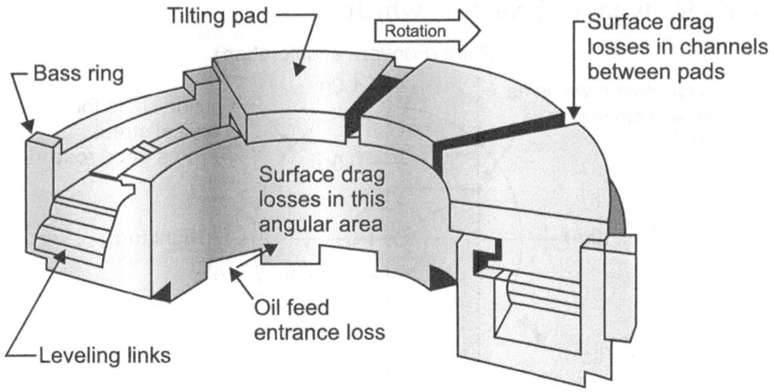


**Fig. 5.19:** Parasitic losses in a pivoted-pad thrust bearing

The load capacity for each bearing is an inverse function of the film thickness cubed. When the load increases to the value at which the pocket pressure reaches the supply pressure, the calculated film thickness.

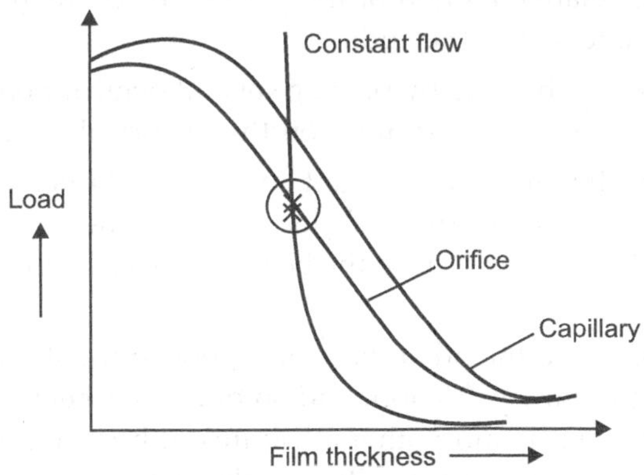


**Fig. 5.20**

## 5.9 ROLLING-CONTACT BEARINGS

In contrast to hydrodynamic bearings, which depend for low-friction characteristics on a fluid film between the journal and the bearing surfaces, rolling-element bearings employ a number of balls or rollers that roll in an annular space. The term rolling element bearing because the contact between the rolling elements and the races or rings consists more of sliding than of actual rolling. A rolling contact implies no interfacial slip; this condition can seldom be maintained, however, because of material deformation and geometric factors. Rolling element bearings ordinarily consist of two races or rings (the inner race and the outer race), a set of rolling elements (either balls or rollers), and a separator (sometimes called a cage or retainer) for keeping the set of rolling elements approximately equally spaced. Rolling-element bearings offer the following advantages when compared with hydrodynamic bearings: (1) low starting friction; (2) low operating friction, comparable to that of hydrodynamic bearings at low speeds and somewhat less at high speeds; (3) less sensitivity to interruptions in lubrication than with hydrodynamic bearings; and (4) capability of supporting combined loads. Rolling-element bearings are more versatile than hydrodynamic bearings, which usually can support only radial or thrust loads. Rolling-element bearings also have disadvantages: (1) they occupy more space in the radial direction than do hydrodynamic bearings and (2) they have a finite fatigue life because of repeated stresses at ball-race contacts points in contrast to hydrodynamic bearings, which:

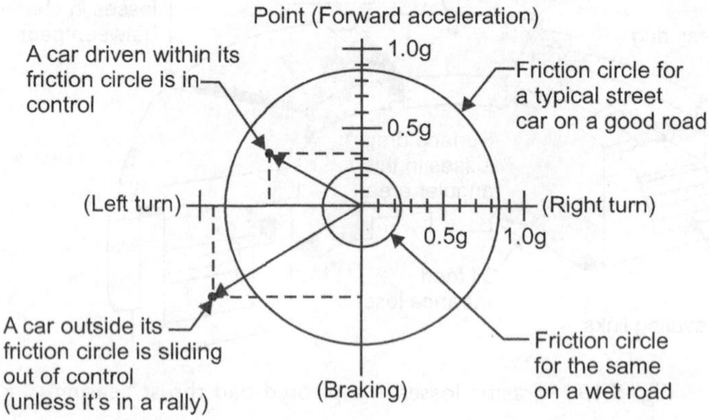


**Fig. 5.21:** A friction circle is a measure of how much grip a car can sustain in any direction

**A rolling-element bearing** is a bearing which carries a load by placing round elements between the two pieces. The relative motion of the pieces causes the round elements to roll with very little rolling resistance and with little sliding.

It is possible to imitate such a bearing by placing several pens or pencils on a table and placing an item on top of them. See "bearings" for more on the historical development of bearings.

A rolling-element rotary bearing uses a shaft in a much larger hole, and cylinders called "rollers" tightly fill the space between the shaft and hole. As the shaft turns, each roller acts as the logs in the above example. However, since the bearing is round, the rollers never fall out from under the load.

Rolling-element bearings have the advantage of a good tradeoff between cost, size, weight, carrying capacity, durability, accuracy, friction, and so on. Other bearing designs are often better on one specific attribute, but worse in most other attributes, although fluid bearings can sometimes simultaneously outperform on carrying capacity, durability, accuracy, friction, rotation rate and sometimes cost. Only plain bearings are used as widely as rolling-element bearings.

## 5.10 FORCES AND STRESSES OF ROLLING-CONTACT BEARINGS

### Load Distribution of Operating Bearings

Force distribution of centripetal bearing with radial loading.

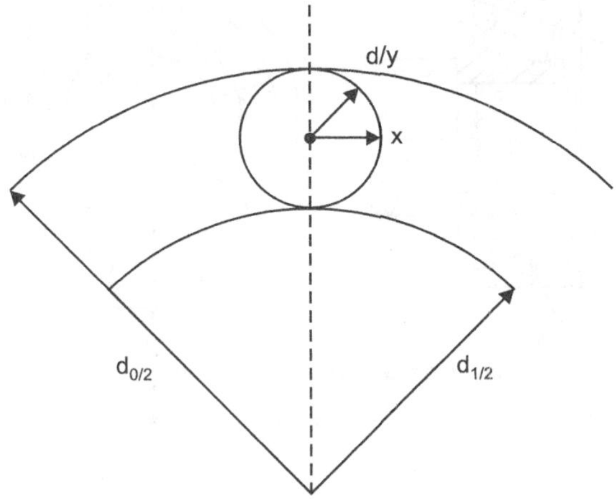


Fig. 5.22

Changes of load and stress

$$d_c = d_f = \frac{1}{2}(d_0 - d_f)$$

$$\therefore \quad d_f = \frac{1}{2}(d_0 + d_1)$$

where (Young's modulus) $$E' = \frac{2}{\dfrac{1-v_1^2}{E_1} + \dfrac{1-v_2^2}{E^2}} = 228 \text{ GPa}$$

$$d_c = \frac{d_0 + d_i}{2} = 0.08 \text{ m}$$

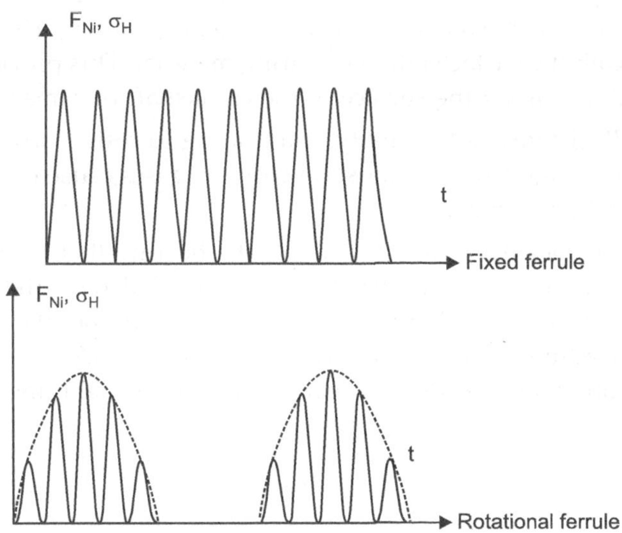


Fig. 5.23

**The Influence of Thrust Load on Load Distribution**

loading on only one roller:

$$F_a = F_d = F_{r\tan\alpha}$$

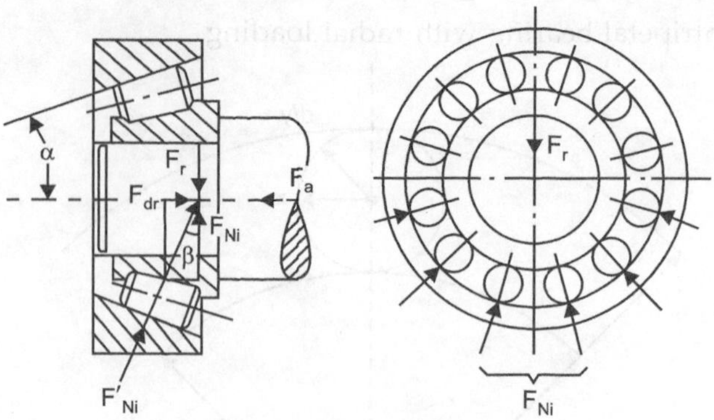


Forces on tapered roller bearing

**Fig. 5.24**

## Bearing Characteristics

Engine bearings must operate under tremendous loads, severe temperature variations, abrasive action, and corrosive surroundings. Essential bearing characteristics include the following:

**Bearing Load Strength** is the ability of a bearing to withstand pounding and crushing during engine operations. The piston and rod can produce several tonne of downward force. The bearing must not fatigue, flatten, or split under these loads. If the bearing load resistance is too low, the beating can smash, fail, and spin in its bore. This ruins the bore or the journal.

**Bearing Conformability** is the ability of a bearing to move, shift, conform to variations in shaft alignment, and adjust to imperfections in the surface of the journal. Usually, a soft metal is placed over hard steel. This lets the bearing conform to the defects in the journal.

**Bearing Embedability** refers to the ability of a bearing to permit foreign particles to become embedded in it. Dirt and metal are sometimes carried into the bearings. The bearing should allow the particles to sink beneath the surface into the bearing material. This prevents the particles from scratching, wearing, and damaging the surface of the crankshaft or camshaft journals.

**Bearing Corrosion Resistance** is the ability of a bearing to resist corrosion from acid, water, and other impurities in the engine oil. Combustion blow-by gases cause engine oil contamination that can also corrode engine bearings.

**Bearing Materials** the backing material (body of the bearing that contacts stationary parts) for engine bearings is normally steel. Softer alloys are bonded over the backing to form the bearing surface. Any one of three basic types of metal alloys can be plated over the top of the steel backing—Babbitt (lead-tin alloy), copper or aluminum (Fig. 5.25). These three metals may be used in different combinations to design bearings for either light-,medium-, or heavy-duty applications.

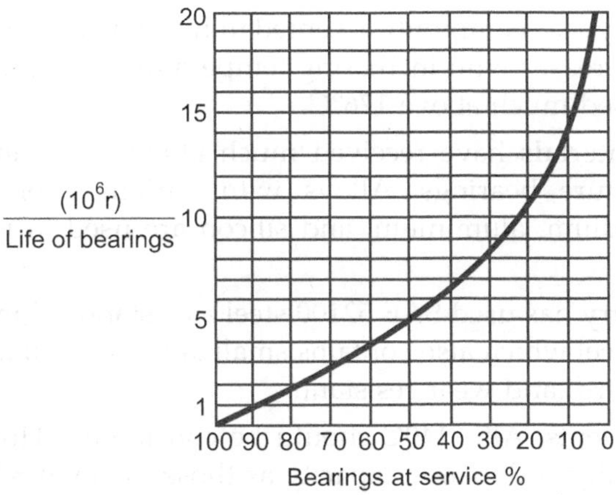


**Fig. 5.25:** Life distribution curve of bearing

**Friction Losses in Rolling-Element Bearings**

Some of the factors that affect the magnitude of friction losses in rolling bearings are:

(1) Bearing size
(2) Bearing type
(3) Bearing design
(4) Load (magnitude and type, either thrust or radial)
(5) Speed
(6) Oil viscosity
(7) Oil flow.

In a specific bearing, friction losses consist of:

(1) Sliding friction losses in the contacts between the rolling elements and the races. These losses include differential slip and slip due to ball spinning. The shearing of elastohydrodynamic lubrication films, which are extremely thin and contain oil whose viscosity is increased by orders of magnitude above its atmospheric pressure value.

(2) Hysteresis losses due to the damping capacity of the race and ball materials.

(3) Sliding friction losses between the separator and its locating race surface and between the separator pockets and the rolling elements.

(4) Shearing of oil films between the bearing parts and oil churning losses caused by excess lubricant within the bearing.

(5) Flinging of oil off the rotating parts of the bearing because of the dominant role played in bearing frictional losses by the method of lubrication, there are no quick and easy formulas for computing rolling element bearing power loss. Coefficients of friction for a particular bearing can vary by a factor of 5 depending on lubrication.

## 5.11 BEARING MATERIALS

Bearings materials are classified as through-hardened materials. Critical applications require vacuum-processed steels.

Regardless of material type, a commonly accepted minimum hardness for bearing components is 58 $R_c$. Since hardness decreases with increasing temperature, conventional bearing steels such as AISI 440°C cannot be used much above 176°C.

**Through-hardened materials** have received much attention in an effort to develop steels suitable for high-temperature bearings. Alloys with various proportions of molybdenum, tungsten, chromium, vanadium, aluminium, and silicon are used to produce hardness at high temperatures.

The ball-bearing industry has used SAE 52100 steel as a standard material since 1920. This is a high-carbon chromium steel which also contains small amounts of manganese and silicon. This air-melted alloy is clean, hard, and wear resistant.

For corrosive environments, AISI 440C should be considered. However, bearings made of AISI 440C do not have as high a dynamic capacity as those made of SAE 52100.

Halmo and M-50 are usable up to about 315°C. M-50 is the more widely used. WB-49 to 540°C. Oxidation resistances of M-1 and M-2 are marginal above 900°F or 480°C.

For M-50 and M-1 the dynamic capacity can be exceeded at 315°C, but dynamic capacity of bearings made of WB-49 should be derated above that temperature.

Generally, bearings made from M-series materials cost about 50% more than those made from SAE 52100, because of grinding difficulties. At temperatures below 175°C, there appears to be no technical advantage of M-series materials over SAE 52100.

**Carburized, or case-hardened materials**, are characterized by a hardened surface (greater than 3.8 mm thick) and a soft core. At room temperatures, the surface hardness of these materials run around $R_c$ 58 to 63, and their core hardnesses are $R_c$ 25 to 48. These materials are generally limited to temperatures less than 175°C.

Where shock and high vibrational loads are present, carburized materials may have an advantage because of their soft, ductile inner core.

Vacuum-melt steels provide increased reliability and basic dynamic load rating because they have a lower inclusion content than air-melted steels. However, the improvement in bearing life is not always commensurate with the improvement in cleanliness.

Induction vacuum melting is a process in which a cold charge is melted in an induction furnace and then poured into ingots; the whole operation is performed in vacuum. Drawbacks of this process are high cost and variations in quality because of refractory problems.

A better way to make bearing steels is consumable-electrode vacuum melting. Electrodes made from air-melted heat are remelted under vacuum by electric arcs. The remelted product solidifies in a water-cooled copper mold under vacuum. This gives different solidification conditions than the induction-melting technique and produces a more consistent high-quality steel.

Chapter 6

# Surface Topography

## 6.1 INTRODUCTION TO TRIBOLOGICAL SYSTEMS

A tribological system consists of the surfaces of two components that are in moving contact with one another and their surroundings. Progress and extent of wear are determined by the materials and finishes of the components, any intermediate materials, surrounding influences and operating conditions.

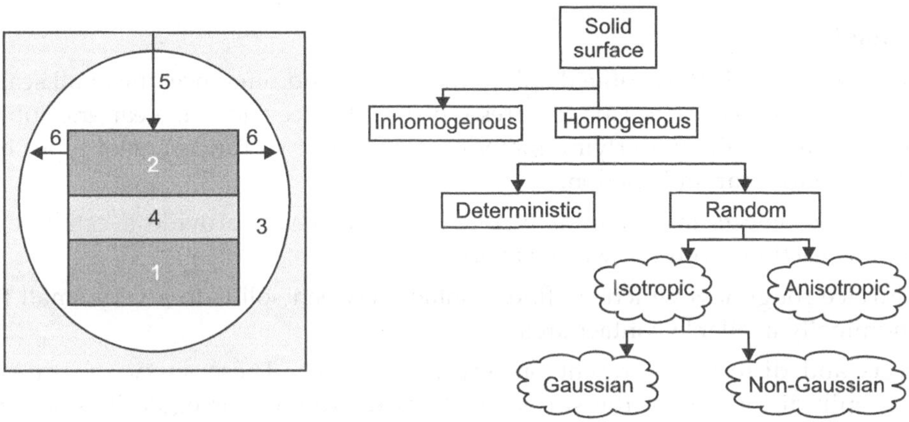


**Fig. 6.1:** Tribological system

- Base object Base structure.
- Opponent body.
- Surrounding influences: Temperature, relative humidity, pressure.
- Intermediate material: Oil, grease, water, Particles, contaminants.
- Load on surfaces.
- Motion displacement.

$$R = CLA = AA = \frac{1}{L}\int_0^L (Z - M)\,dx$$

$$\therefore \quad m = \frac{1}{L}\int_0^L z\,dx$$

$$\sigma^2 = \frac{1}{L}\int_0^L (z - m)^2\,dx$$

$$\therefore \quad R_q = \sigma$$

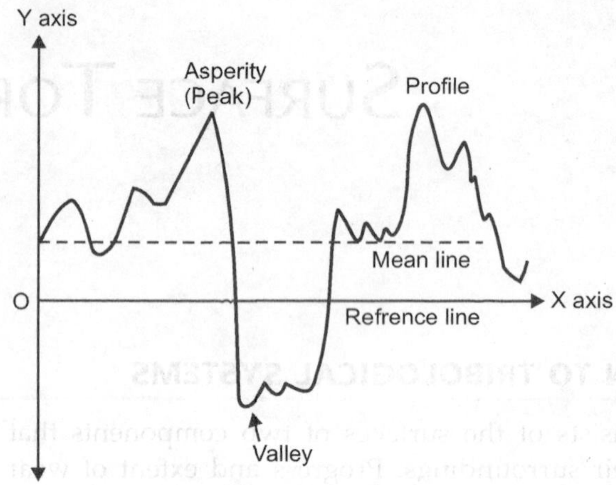


**Fig. 6.2**

## Surfaces of Solids

On the surfaces of the solids the profile displays the valleys and reference line at all scales of size, surfaces of solids contain characteristic features which influence friction, wear and lubrication in a manner independent of the underlying material. There are two fundamental types of features of special relevance to wear and friction:

- atomic-scale defects in a nominally plain surface which provide a catalytic effect for lubricant reactions with the worn surface.
- the surface roughness which confines contact between solids to a very small fraction of the nominally available contact area.

The defects and distortions present on any real surface. These surface features, ranging particles from bulk distortions of the surface to local microscopic irregularities, exert a strong influence on friction and wear. The imperfections and features of a real surface influence the chemical reactions or lubricants while the visible roughness of most surfaces controls the mechanics of contact between the solids and the resulting wear. The atoms of the solid body can be visualized as hard spheres packed together with no loose space. The atom layers usually lie inclined to the surface. As a result a series of terraces is formed on the surface generating a quasi-planar surface. Atoms present on the surface have a lower bonding strength than interior atoms because they have a lower number of adjacent atoms. It has been observed that without all of these imperfections surfaces would probably be virtually inert to all chemical reactants.

Chemical reactions between the surface and the lubricant. The reaction between lubricant and surface often produces a surface layer or 'film' which reduces friction and wear. Furthermore, the substrate material may be deformed plastically, which increases the number of dislocations reaching the surface. Dislocations form strong catalytic sites for chemical reactions and this effect is known as 'mechanical activation'. An intense plastic deformation at a worn surface is quite common during wear and friction and the consequent mechanical activation can exert a strong influence on the formation of a lubricating film.

$$A_r = \eta\pi A\,(2.3\beta)^2 \int_0^m (z-d) \int_0^m f\,\frac{(z, C)}{NC}\,dc\,dz$$

$$\therefore \quad r = \frac{2\pi^{0.5}\,(2.3\beta)^2}{9\sigma}$$

$$\therefore \quad w = \frac{4}{3} A\,\sigma\,E\,(2.3\beta) \int_0^m (z-d)^{1.5} \int_0^m \frac{f(z, C)}{N\sqrt{C}}\,dc\,dz$$

## 6.2 SURFACE TOPOGRAPHY

Surface topography is the study of the surface lay where roughness height and waviness components of surface imperfections at an atomic level are matched by macroscopic deviations from flatness. Apart from the cleaved faces of mica , are rough. Roughness means that most parts of a surface are not flat but form either a peak or a valley. Amplitude between the peaks and valleys for engineering surfaces is about one micrometre. The profile of a rough surface is almost random unless some regular features have been deliberately introduced.

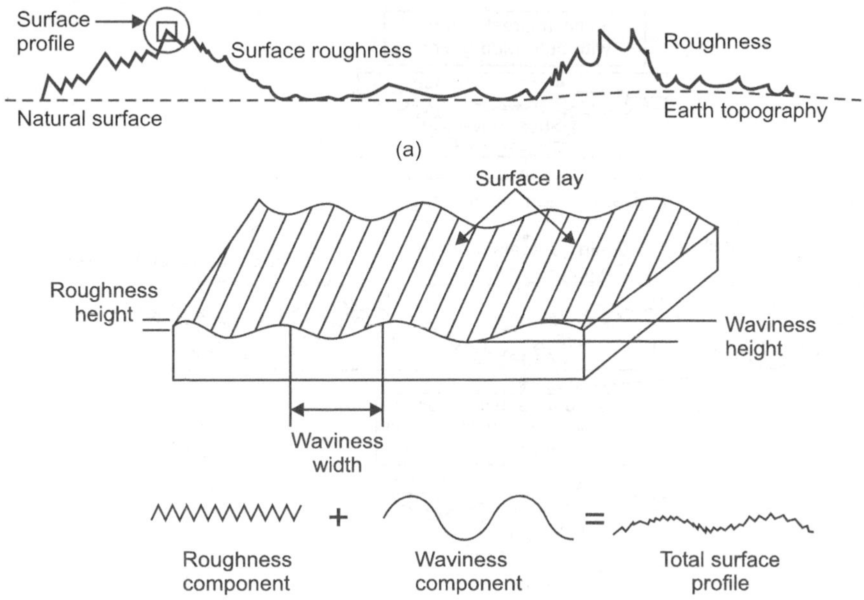


**Fig. 6.3** (a), (b)

The random components of the surface profiles look very much the same whatever their source, irrespectively of the absolute scale of size involved. Property of surface roughness is that, if repeatedly magnified, increasing details of surface features are observed down to the nanoscales. The surface profiles is the same regardless of the magnification where as shown in Fig. 6.4 the load is applied on the surface the actual contact pressure between the asperities and the surface roughness plays a major role on the surface profilemeter/earth topography.

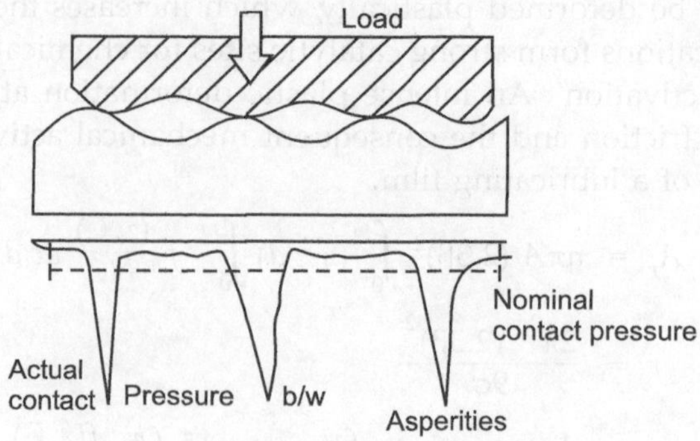


Fig. 6.4

The introduction of statistical methods to the analysis of surface topography was probably due when a bearing area curve as a means of profile representation. This curve representing the real contact area, also known as the Abbott curve, is obtained from the surface profile. It is compiled by considering the fraction of surface profile intersected by an infinitesimally thin plane positioned above a datum plane.

The procedure is repeated through a number of slices. It should also be realized that most real engineering surfaces consist of a blend of random and non-random features. The series of grooves formed by a shaper on a metal surface are a prime example of non-random topographical characteristics. On the other hand, bead-blasted surfaces consist almost entirely of random features because of the random nature of this process.

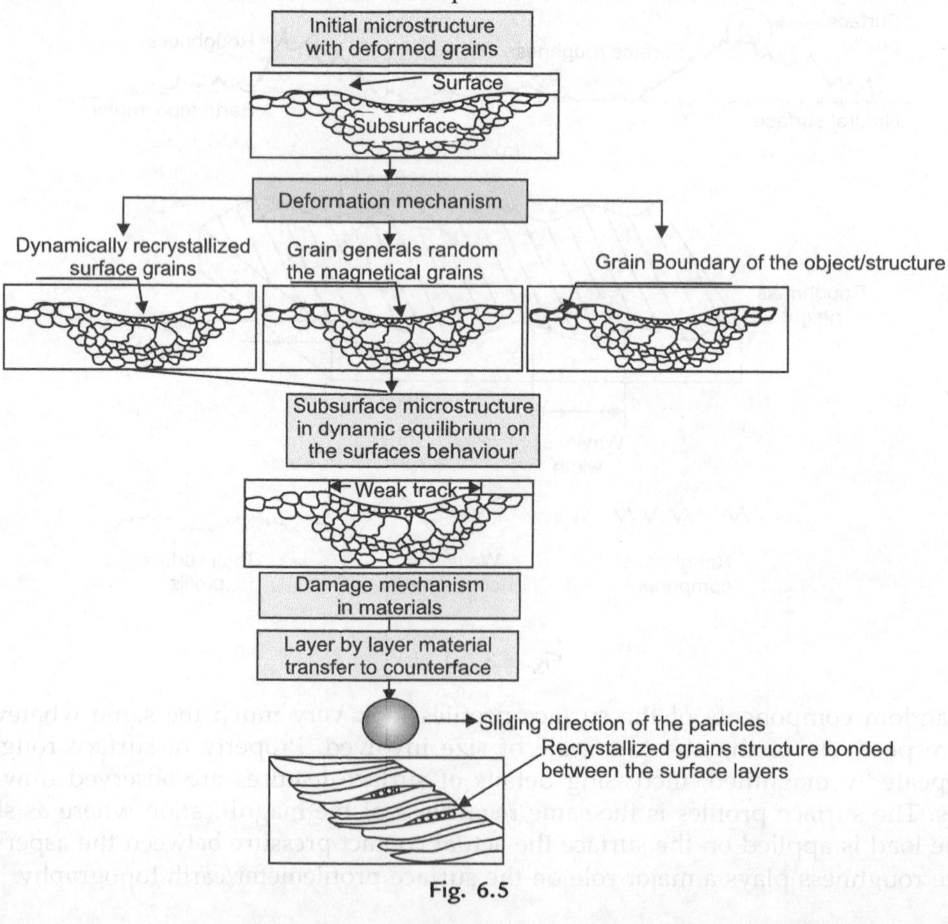


Fig. 6.5

**Characterization of Surface Topography**

The most widely used surface descriptors are the statistical surface parameters. A new development in this area involves surface characterization by fractals.

- *Characterization of Surface Topography by Statistical Parameters*:

Real surfaces are difficult to define. Surface at least two parameters are needed, one describing the variation in height (*i.e.*, height parameter), the other describing how height varies in the plane of the surface (*i.e.*, spatial parameter). The deviation of a surface from its mean plane is assumed to be a random process can be described using a number of statistical parameters. Height characteristics are commonly described by parameters such as the centre-line-average or roughness average root mean square roughness (***RMS*** or ***'Rq'***), mean value of the maximum peak-to-valley height (***'Rtm'***), ten-point height (***'Rz'***) and many others. The ***'Ra'*** represents the average roughness over the sampling length. The effect of a single spurious, non-typical peak or valley (*e.g.*, a scratch) is averaged out and has only a small effect on the final value. In Fig. 6.6 the soft material, extra depth and surface roughness is shown.

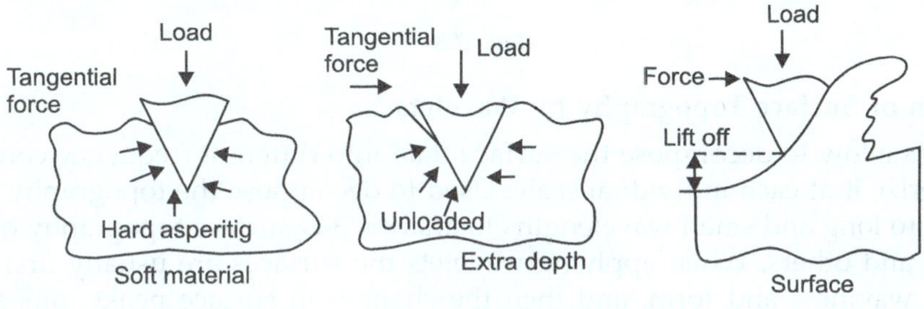


**Fig. 6.6**

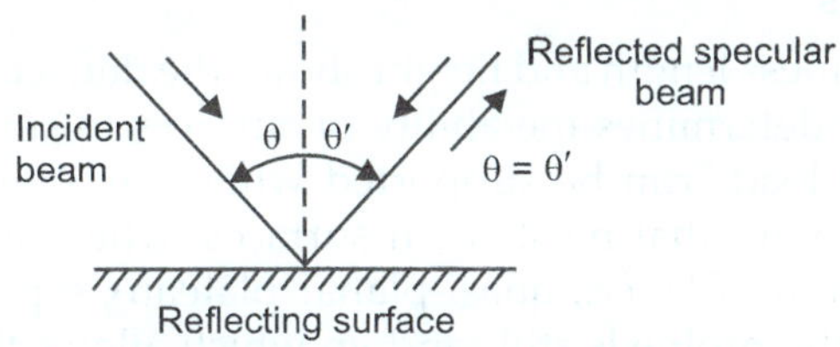


**Fig. 6.7**

Real surfaces can be described by a number of statistical functions. Two surfaces can have the same height parameters their spatial arrangement and hence their wear structure function (**SF**) or the power spectral density function (**PSDF**) are commonly used. These functions are used to discriminate between the differing spatial surface characteristics by examining their decaying properties.

Engineering surfaces is that they exhibit topographical details over a wide range of scales; from nano-to micro-scales. Surface topography is a nonstationary random process for which the variance of height distribution ($RMS_2$) depends on the sampling length. Therefore the same surface can exhibit different values of the statistical parameters when a different sampling length or an instrument with a different resolution is used. This leads to certain inconsistencies in surface characterization. The main problem is associated with the discrepancy between the large number of length scales that a rough surface contains and the small number of particular length scales, *i.e.*, sampling length and instrument resolution, that are used to define the surface parameters.

**Characterization of Surface Topography by Fourier Transform**

Fourier transform methods allow to decompose the surface data into complex exponential functions of different frequencies. To calculate the power spectrum and the autocorrelation function in order to obtain the surface topography parameters. The problem is that they provide results which strongly depend on the scale at which they are calculated, and hence they are not unique for a particular surface.

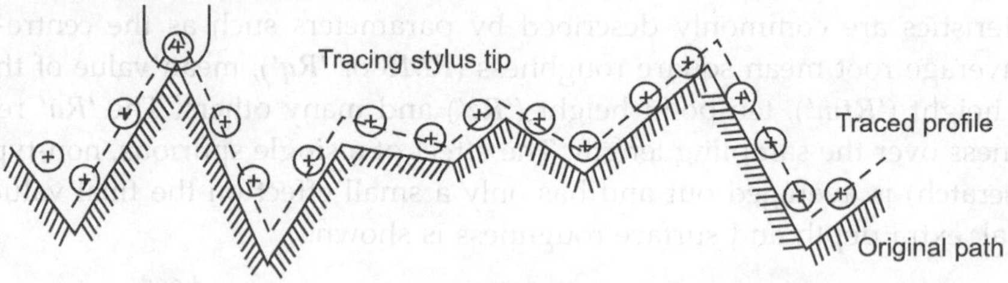


**Fig. 6.8**

**Characterization of Surface Topography by Wavelets**

Wavelet methods allow to decompose the surface data into different frequency components and then to characterize it at each individual scale. Used to decompose the topography of a grinding wheel surface into long and small wavelengths to analyse 3-D surface topography of orthopaedic joint prostheses and others. When applying wavelets the surfaces are usually first decomposed into roughness, waviness and form, and then the changes in surface peaks, pits and scratches together with their locations are obtained at different scales.

**Optimum Surface Roughness**

Surface roughness where waviness length and height shows the deflection of the surface roughness of components is critical as it determines the ability of surfaces to support load. At high or very low values of ***'Rq'*** only light loads can be supported while the intermediate ***'Rq'*** values allow for much higher loads. It is found that most worn surfaces, where lubrication is effective, tend to exhibit a favourable surface profile, *i.e.*, quasi-planar plateaux separated by randomly spaced narrow grooves. In this case, the profile is still random which allows the same analysis of contact between rough surfaces.

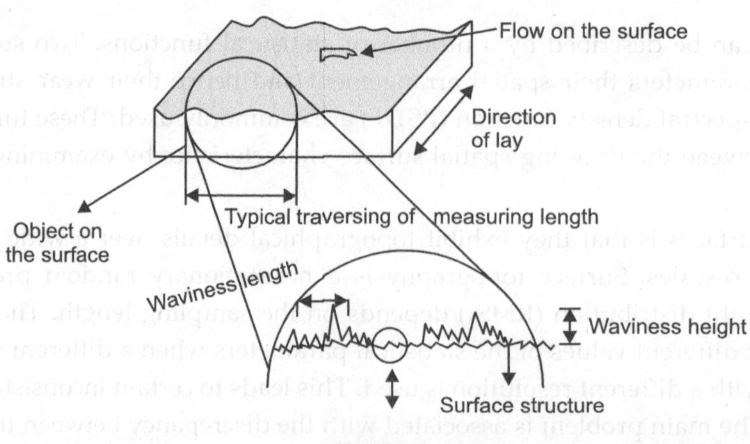


**Fig. 6.9**

### Effect of Sliding on Contact Between Solid Surfaces

Analyses of contact between solids are based on stationary contacts where no sliding occurs between the surfaces. Parameters such as the real contact area and the average contact stress under sliding are of critical importance to the interpretation of wear and friction so that analysis of solid to solid contact under sliding.

A qualitative description of contacts between asperities during sliding has been obtained from studies of hard asperities indenting a soft material. Three distinct stages of contact were observed: static contact, *i.e.*, where the tangential force is small, the stage just before the gross movement of the asperity, *i.e.*, when the tangential force is at its maximum level, and unrestricted movement of the asperity. Tangential force is equivalent to frictional force in real sliding contacts.

### Techniques of Surface Treatment

| | |
|---|---|
| Physical and chemical structure | Thin discrete coating; no limitations on materials vapour deposition on the surface particles. |
| Ion implantation | Thin diffuse coating; mixing with substrat inevitable particles. |
| Laser glazing and alloying | Thick coatings; coating material must be able melt. |
| Electroplating | Wide range of coating thicknesses, but adhesion to substrate is poor technique and only certain materials can be coated by this techniques. |
| Friction surfacing | Simple technology but limited to planar surfaces produces thick metal coating. |
| Explosive cladding | Rapid coating of large areas possible and bonding to substrate is good. Can give a tougher and thicker coating than many other methods. |
| Thermal spraying | Very thick coatings possible but control of coating purity is difficult because the thermal behaviour. |
| Surface welding | Suitable for very thick coatings only; limited to materials stable at high temperatures; coated surfaces may need further preparation. |

## 6.3 SURFACE EXAMINATION TECHNIQUES

### Introduction

Non-Destructive Testing (NDT) is a form of testing or inspection of the structural integrity of a component without compromising its ability to perform in service by testing the surfaces.

It has already become or is rapidly becoming mandatory for safety critical components. In applications such as aerospace, oil and gas, pressure vessels and nuclear power, for the verification of weld quality or the monitoring of corrosion damage in service. These are used to examine the particles and surfaces by the non-destructive tests have to be non-damaging and non-invasive, and methods and techniques have been developed to considerable levels of sophistication. They all require operation of equipment and interpretation of results by skilled, well trained personnel. How the D.T and NDT testing techniques now a day very helpfull to find out the defects between the particle to particle contact on and between the surface meeting parts. These are various types of find out or examine these defects:

Surface examination and condition monitoring techniques include:

- Visual Inspection
- Liquid Penetrant Testing
- Magnetic Particle inspection

Others are methods used to examine the bulk or whole section of a component or structure include:

- Radiography
- Ultrasonic Inspection
- Acoustic Emission
- Leak Testing

The selection of the most appropriate NDT method for a particular component depends on a number of factors such as:

- Material of construction
- Thickness of section
- Type of defect to be detected
- Cost Inspection equipment available
- Operator skill Level of inspection

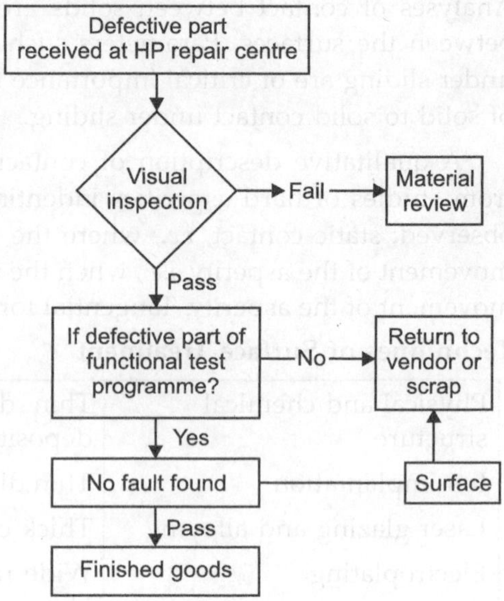


Fig. 6.10

### Visual Inspection

Visual Inspection is probably the first and most widely practiced method of NDT in which particles involves inspecting, directly or indirectly, for surface condition, defects, roughness or dimensional changes and verified the correctly by visualisation the object.

### Applications

Typical applications include:

*Weld Inspection*

- verification of correct weld preparation and defects
- inspection for surface defects, *e.g.*, undercuts and cracks, Beam penetration.
- verification of completed weld profile (cap, root)

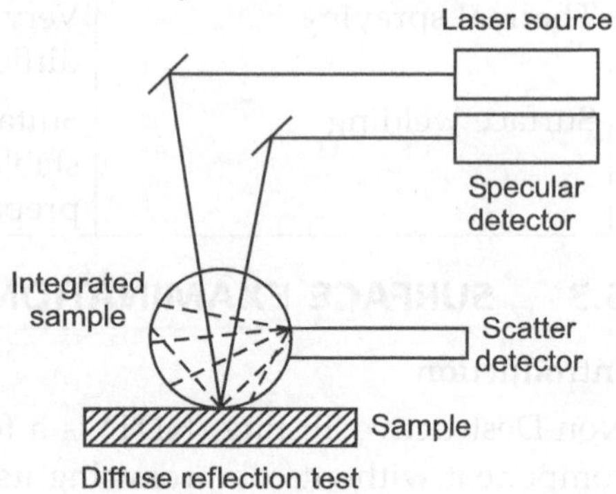


Fig. 6.11

*Metrology*

- measurement of component geometry
- measurement of surface roughness

*Condition monitoring in service*

- detection of corrosion
- detection of defects and cracks

*Etch inspection*

- detection of grinding defects
- detection of incorrect heat treatment (case depth)
- metallographic examination (weld profile, inclusions grain size, microstructure)

*Visual inspection has many techniques to aid the operator and extend its capability.*

- Lighting must be at an appropriate level and of the correct type.
- Magnifiers, from low power, 2x to 5x, hand held lenses to high power microscopes.
- Borescopes are devices that permit inspection of internal surfaces in piping, engines, air frames etc. They contain systems of lenses, mirrors, prisms and lighting which illuminates the test area and transmits images to the inspector. They may be either rigid or flexible. The flexible type uses optical fibres, the smallest have outside diameters of just a few millimetres and can be steered into position around tight bends.
- Television cameras combined with borescopes, video recorders and image processing equipment permit the remote collection and storage of visual images. Image enhancement by computer is also possible.
- Surface replication using strippable films also permits the remote inspection of surfaces.
- Etching for metallographic examination and to reveal cracks, inclusions and localised variations in composition and hardness.

Ultimately, visual testing is dependant upon the eyesight of the inspector. Regular eyesight tests for both visual acuity and colour differentiation are essential. The time for continuous viewing must be restricted with frequent rests to reduce fatigue.

## Liquid Penetrant Testing

Visual inspection this is probably the oldest and most widely used of all the NDT methods. It can be used on any non-porous material. Its use is confined to the detection of surface breaking defects.

A coloured or fluorescent dye, when applied to a clean surface and allowed to dwell for 10 to 20 minutes, will be drawn into the discontinuity by capillary pressure. The capillary pressure is determined by the width of discontinuity, surface tension and contact angle of the dye on the surface.

The excess dye is then removed from the surface and a developer applied, Fig. 6.12.

The developer has two functions to assist in drawing the dye back out to the surface, giving an indication to assist in seeing this indication by giving good visual contrast.

After a time, usually a minimum of 10 minutes, the surfaces are visually inspected. White light is used for red dyes, ultra-violet light for fluorescent dyes.

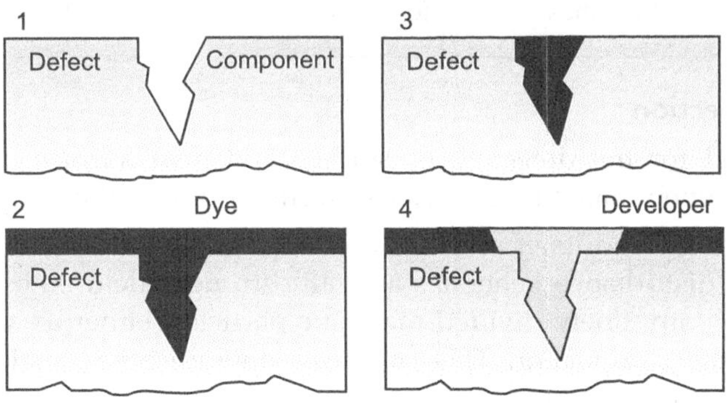


**Fig. 6.12:** Liquid penetrant testing

They are not suitable for components with rough or poor surface finishes as it is difficult to remove the excess dye, causing an overall fluorescence that masks discontinuities, preventing their detection.

They do however make seeing indications in internal bores, cored holes, etc. easier.

## Types of Liquid Penetrant Testing

Liquid Penetrant testing systems are usually classified by the method used for excess dye removal.

There are three main types:

- *Solvent*: Usually supplied in aerosol cans and applied to the surface, excess dye is wiped away with a lint free cloth.
- *Water washable*: Some dyes contain an additive which emulsifies the oil in the dye making it soluble in water.
- *Post emulsification*: The emulsifier is applied just prior to the water wash. Emulsification stage is critical, too short a time means that all excess dye will not be removed causing contrast reduction, too long a time and the dye will be removed from the discontinuities making detection more difficult.

The developer is normally a dry powder carried in a non-aqueous solvent.

| | Part Size | Part Type | Part Shape | Batch Size | Sensitivity | Cost |
|---|---|---|---|---|---|---|
| Solvent Solution | Any, ideal for large items | Any, with reasonable surface finish, usually welds. | Any, in particular external inspection | Small as it is time consuming | Good | Fairly expensive |
| Water washable | Usually small | Good on rough surfaces e.g., castings and forgings | Any, provided easy to wash | Any, ideal for automation | Poor Rough | Cheapest |
| Post Emulsification Solid particles | Small and Micro parts | Precision machined or other good finishes | Can be complex with internal features | Small | Excellent | Expensive, usually needs UV system |

## Magnetic Particle Inspection

A very efficient method for the detection of surface and near surface discontinuities in any ferromagnetic material when magnetic particles movement due to the eddy current flows.

A magnetic field will spread itself uniformly through a component unless disturbed by the presence of a flaw. The disturbance generates a locally stronger field, known as a leakage field. This will strongly attract any finely divided magnetic particles, either as an ink or dry powder, which are applied to the component. This produces an indication, making the flaw visible, as shown in Fig. 6.13.

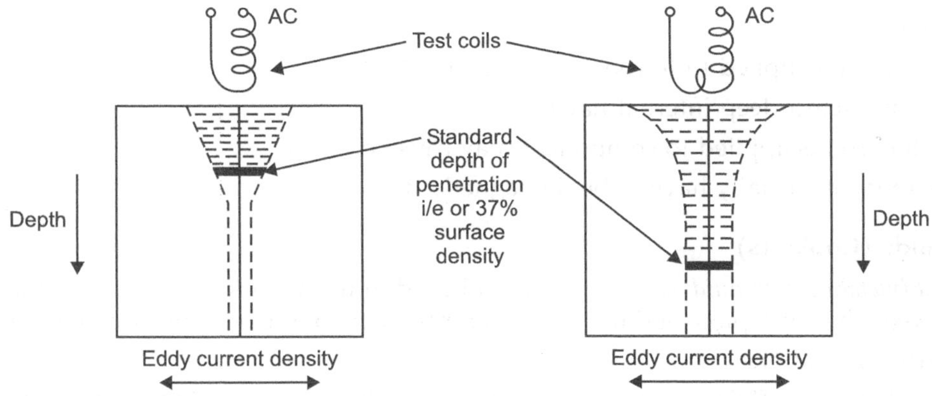


**Fig. 6.13**

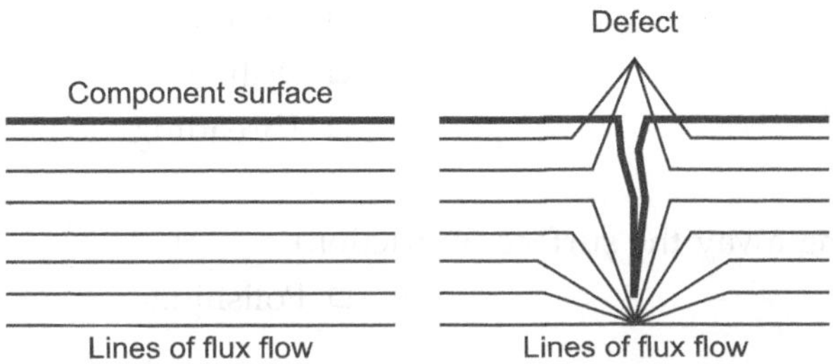


**Fig. 6.14:** Magnetic particle inspection

Factors can influence the production and quality of indications defect size depth of defect below surface magnetising relative orientation of defect and direction of magnetisation magnetic properties of test piece size and magnetic properties of detecting media.

The orientation of the flaw with respect to the direction of the flow of magnetic flux greatly influences the likelihood of creating a leakage field. Ideally the flaw should be at 90° to the flux flow, but detection may still be possible at 45°.

Magnetic flux can be induced in a test piece by using a permanent or electromagnet passing an electrical current through the test piece passing an electric current through a coil which surrounds the test piece and the magnetic field produces current between surfaces.

When dealing with large components, the simplest means of inducing a magnetic field is with a permanent or electromagnet. It is normal practice to paint the test piece white and to use black magnetic ink particles suspended in paraffin for the greatest contrast, again two tests are required at 90 to each other for maximum certainty of detection of flaws.

The other methods are better suited to routine testing in laboratories or inspection departments. The limiting factor is now the 'bed size', *i.e.*, the length between the head stocks which hold a component and allow the magnetising current to flow.

**Advantages**

- Simple and easy to conduct
- Detect surface and near surface flaws
- Can detect flaws filled with contaminants e.g. oxide or non metallic inclusions
- Sensitivity of testing can be specified and checked

**Disadvantages**

- Can only be applied to ferromagnetic materials
- Will not detect deep internal flaws
- High currents applied to component may cause damage
- Components usually have to be demagnetised.

## Cutting Fluids (Coolants)

*Cutting fluid (metalworking fluid, coolant)* is a liquid used in metalworking operations for reducing friction between the work piece and the tool and for removal of the heat generated by the friction.

**Metalworking Operations**

Cutting (metalworking) fluids are involved in the following metalworking operations:

- Cutting (separation of metal from a work piece in the form of chips)
  - ❑ Milling
  - ❑ Turning
  - ❑ Boring
  - ❑ Drilling
  - ❑ Broaching
  - ❑ Threading
  - ❑ Sawing
- Abrading (rubbing away the surface by friction)
  - ❑ Grinding
  - ❑ Polishing
  - ❑ Lapping
- Metal forming (changing the shape of a work piece by a pressure)
  - ❑ Rolling
  - ❑ Forging
  - ❑ Extrusion
  - ❑ Drawing
  - ❑ Deep drawing
  - ❑ Stamping

## Benefits of Metalworking Fluids (Coolants)

*Lubrication and cooling effects provided by cutting fluids (coolants) result in the following benefits of metalworking*:

- **Better surface finish and smooth surface.**
  Cutting (metalworking) fluids lubricate the workpiece-tool metal-to-metal contact zone preventing tool galling and seizure, which assures good surface finish.
- **Longer tool life and high performance.**
  Temperature of a tool properly cooled by a cutting fluid (coolant) does not exceed the critical value, beyond which the tool hardness drops and its wear rate increases.
- **Narrower tolerances of the work piece size.**
  Cutting fluids provide both: decreasing the amount of heat generated in the workpiece-tool contact zone due to reduction of the friction and removing the heat. Thus cutting fluid stabilizes the work piece temperature providing better control of its geometry. Cutting fluids reduce the tool wear and prevent formation of built-up edges.
- **Lower energy consumption and reduced friction.**
  Reduced friction provided by a metalworking fluid results in decreased work of the operation (cutting, abrading or forming).

- **Cleaner cutting zone.**

  Cutting fluids remove the chips and fines formed in cutting (abrading) operations keeping the cutting zone clean and preventing the surface damage.

- **Better corrosion protection.**

  Metalworking (cutting) fluids may contain Corrosion inhibitors. Such fluids form a protection film on the work piece surface, machine parts and chips

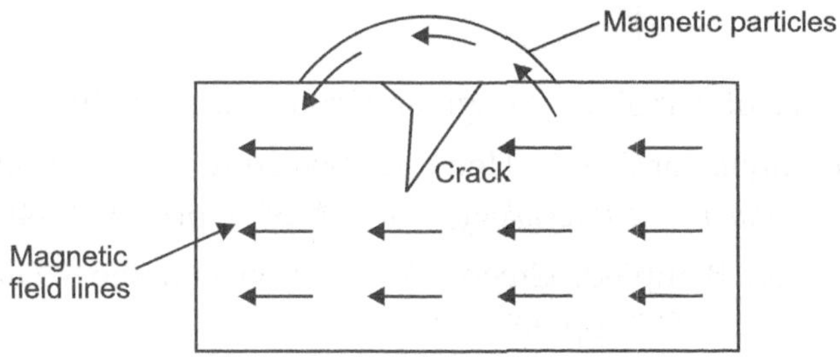


Fig. 6.15

## Deterministic vs. Probabilistic (Stochastic)

A **mathematical model** is a description of a system using mathematical language. The process of developing a mathematical model is termed **mathematical modelling.** Mathematical models are used not only in the natural sciences and engineering disciplines but also in the social sciences. Physicists, engineers, statisticians, operations research analysts and economists use mathematical models most extensively.

Mathematical models can take many forms, including but not limited to dynamical systems, statistical models, differential equations, or game theoretic models. These and other types of models can overlap, with a given model involving a variety of abstract structures.

A deterministic model is one in which every set of variable states is uniquely determined by parameters in the model and by sets of previous states of these variables. Perform the same way for a given set of initial conditions. Conversely, in a stochastic model, randomness is present, and variable states are not described by unique values, but rather by probability distributions.

Deterministic models describe behaviour on the basis of some physical law. For example, the planets move around the sun according to Newton's laws and their position can be predicted with great accuracy well into the future. Deterministic relationship is unlikely due to unpredictable factors—for example, a comet moving through the solar system could collide with a planet or a moon and perturb its orbit. Newton's laws can only predict the positions of two bodies, such as the sun and the earth exactly. If there is a third body, such as the moon, there is no longer a simple mathematical that can describe and predict the paths of all three with unlimited accuracy for all time. When applied to the whole solar system, even supercomputers can't use Newton's laws to predict the resulting chaos.

Where the influence of several unknown factors is sizable, exact prediction is not possible, but it may be possible to predict to within a known confidence interval—or to predict the probability that a particular value will be observed at a particular time. This is called a stochastic (or probabilistic) process.

Many "agents" act independently to influence the economy. While we know there is a relationship between interest rates, say, and economic growth the relationship is not deterministic and there is also "noise" caused by the random actions of the many agents.

# REFERENCES


1. D. Dowson, History of Tribology, Longman Group Limited, 1979.
2. S. Adelman, C.R. Taylor and N.C. Heglund, Sweating on Paws and Palms: What is Its Function, *American Journal of Physiology*, Vol. 29, 1975, pp. 1400–1402.
3. N.H. Sleep and M.L. Blanpied, Creep, Compaction and the Weak Rheology of Major Faults, *Nature*, Vol. 359, 1992, pp. 687–692.
4. B.W. Kelley, Lubrication of Concentrated Contacts, Interdisciplinary Approach to the Lubrication of Concentrated Contacts, Troy, New York, NASA SP-237, 1969, pp. 1-26.
5. P.A. Pacholke and K.M. Marshek, Improved Worm Gear Performance with Colloidal Molybdenum Disulfide Containing Lubricants, *Lubrication Engineering*, Vol. 43, 1986, pp. 623–628.
6. Lubrication (Tribology)-Education and Research. A Report on the Present Position and Industry Needs, (Jost Report), Department of Education and Science, HM Stationary Office, London, 1966.
7. Research Report (T76-38) Tribologie (Code BMFT-FBT76-38), Bundesministerium Fur Forschung and Technologie (Federal Ministry for Research and Technology), West Germany, 1976.
8. Strategy for Energy Conservation Through Tribology, ASME, New York, November, 1977.
9. Engineering Tribology Gwidon W. Stachowiak Department of Mechanical and Materials. Engineering,University of Western Australia, Australia and Andrew W. Batchelor Department of Mechanical and Materials Engineering, University of Western Australia, Australia.
10. Friction Science and Technology from concepts to applications by Peter J. Blau.
11. Solid Lubricants and Coatings for Extreme Environments: State-of-the-Art Survey Kazuhisa Miyoshi Glenn Research Center, Cleveland, Ohio.
12. Ausman, J. S. 1959. Theory and design of self-acting gas-lubricated journal bearings including misalignment effects. Proc. 1st Int. Symp. Gas-Lub. Bearings, ACR-49, 161-192, ONR Washington.
13. D.C.Ausman, J. S. 1961. An improved analytical solution for self-acting, gas lubricated journal bearings of finite length. ASME J. Basic Eng., 83, 188-194.

14. Ausman, J. S. 1966. Gas-lubricated bearings. Advanced Bearing Technology, E. E. Bisson and W. J. Anderson (ed.). NASA, Washington, D.C.

15. Burgdorfer, A. 1959. The influence of molecular mean free path on the performance of hydrodynamic gas lubricated bearings. ASME J. Basic Eng., 81, 94-100. Chan, D. Y. C. and Horn, G. 1985. J. Chemical Physics, 83, 5311.

16. Harrison, W. J. 1913. The hydrodynamical theory of lubrication with special reference to air as a lubricant. Trans. Cambridge Phil. Soc, 22, 39-54.

17. Him, G. A. 1984. Sur les principaux phenomenes Bull. Soc. Ind. Mulhouse, 26, 188. Hsia, Y.-T. and Domoto, G. A. 1983. An experimental investigation of molecular rarefaction effects in gas lubricated bearings at ultra-low clearances. ASME Journal of Lubrication Technology, 105, 120-130. Kingsbury, A. 1897. Experiments with an air lubricated journal. /. Am. Soc. Naval Engineers, 9, 267-292.

18. Mitsuya, Y. 1993. Modified Reynolds equation for ultra-thin film gas lubrication using 1.5-order slip-flow model and considering surface accommodation coefficient. ASME Journal of Tribology, 115, 289-294. Nayfeh, A. H. 1973. Perturbation Methods. Wiley, New York.

19. Pan, C. H. T. 1981. Gas bearings. In Tribology: Friction, Lubrication and Wear, A. Z. Szeri (ed.). McGraw-Hill, New York. Pinkus O. and Sternlicht, B. 1961. Theory of Hydrodynamic Lubrication. McGraw- Hill, New York. Raimondi, A. A. 1961. A numerical solution of the gas lubricated full journal bearing. ASME Trans.

20. Wear – Materials, Mechanism and practice Edited by Gwidon W. Stachowiak.

21. Barwell, F.T., 'Wear of Machine Elements', Proceedings of the International Conference on Fundamentals of Tribology, MIT Press, Boston, Massachusetts, 1978, pp. 401–440.

22. Landsdown, A.R. and Price, A.L., 'Materials to Resist Wear', Pergammon Press, New York, 1986.

23. Jones, M.H. (ed), Proceedings of the International Conference on Ferrography, University of Wales Swansea, UK, 1982.

24. Ballard, D.H. and Brown, C.M., 'Computer Vision', Prentice-Hall, Enylewood Cliffs, NJ, 1982.

25. Boundary Lubrication and Lubricants by Seiichro Hironaka School of Engineering, Tokyo Institute of Technology.

26. F:\tribology mera\Fig1-1.gif, F:\tribology mera\pw1502071.gif, WWW.google.com, images.in.

Theses References Papers and Books guide me about the Tribology subject so that with the help of these article/google images, I understand the topics wear, friction, lubrication, bearings or surface topography and work on Tribology Concept. I am thankful to those whose help me to write my concept about Tribology, Specially these Authors and References Papers are very helpful for me.

**BT-6/J04**

**MET–304 : Tribology**

**Time : 3 Hours** **Maximum Marks : 75**

*Note:* (1) *Attempt any five questions.*

(2) *All questions carry equal marks.*

(1) What are tribological systems? What are their characteristic features? Discuss analysis and assessment of surface topography. 15

(2) What are quantitative laws of sliding friction? Discuss adhesive theory of friction. Also discuss the criticism of adhesive theory. 15

(3) Discuss abrasive wear. What are various types of abrasive wear? Give some applications of abrasive wear. What are the effects of hardness, particles size and moisture contents on abrasive wear? 15

(4) Explain various types of wear in detail. How the wear can be prevented? 15

(5) Explain hydrostatic end thrust bearings. What are the required properties of lubricants? What are lubricant additives and why they are used? 15

(6) Explain journal bearings with specialized applications. What are the general requirements and types of bearing materials? 15

(7) What is a porous bearings? For what purpose, the porous bearings are used. What is the material used for porous bearings. 15

(8) Sketch various types of ball bearings and explain the working of each type. Why these bearings are known as antifriction bearings? 15

## BT-6/M06

## Tribology (2005–06)

## Paper : ME-304E

**Time : 3 Hours** **Maximum Marks : 100**

**Note:** Attempt and **FIVE** questions, at least **ONE** from each unit.

### UNIT–I

1. (*a*) Explain Tribological System. Give analysis and assessment of surface topography. 10

   (*b*) What are the causes of friction? State and explain adhesion theory of friction with suitable diagram. 10

2. (*a*) What are the different techniques of surface examination? 10

   (*b*) What are the sources of wear? Explain stick-slip and molecular attraction theory with neat diagram. 10

### UNIT–II

3. (*a*) Explain abrasive wear and classify. Explain cavitation with neat sketch. 10

   (*b*) Write short note on:

   (*i*) Frettling wear (*ii*) Frettling fatigue

   (*iii*) Stick Regime (*ii*) Mechanism of wear. 10

4. (*a*) Write short note on following with examples:

   (*i*) Gas carried solid particle erosion.

   (*ii*) Liquid carried solid particle erosion.

   (*iii*) Liquid impingement erosion.

   (*iv*) Corrosive wear. 10

   (*b*) Explain relative importance of wear. What are the different methods for prevention of wear? 10

### UNIT–III

5. (*a*) Explain Boundary and elasto-hydrodynamic lubrication with neat sketch. 10

   (*b*) Explain different properties of Lubricants and also state the applications of lubricants. 10

6. (*a*) Explain in detail Hydrostatic and Hydrodynamic lubrication with neat sketch. 10

   (*b*) What are the different additives used in lubricating oil and functions of lubrication? 10

### UNIT–IV

7. (*a*) Explain in detail the circular step thrust bearing and write down the pressure distribution equation. 10

   (*b*) What are the different types and configuration of hydrostatic bearing? Explain. 10

8. (*a*) Explain hydrodynamic Gas bearing and give the advantages and disadvantages of a porous bearing. 10

   (*b*) State and prove the pressure equation of a shoit bearing? What are the different bearing materials used? Explain. 10

**BT-6/M05**

**Tribology**

**Paper : MET-304**

**Time : 3 Hours** **Maximum Marks : 75**

**Note:** Attempt any **FIVE** questions.

1. Explain tribological systems and their characteristic features. Give analysis and assessment of surface topography. 15

2. What are the causes of friction? What is adhesion theory of friction? Explain quantitative laws of friction. 15

3. Explain various types of wear. Discuss in detail the abrasive wear. How the abrasive wear can be prevented? Discuss abrasive wear resistant materials. 15

4. (*a*) Explain the relative importance of various types of wear. 6

   (*b*) Define erosion wear, fretting wear and cavitation. Explain with examples. 9

5. Explain hydrostatic and elasto-hydrodynamic lubrication. What is the function of lubricant? Discuss desirable properties of lubricants. 15

6. With the help of a diagram discuss hydrostatic bearings. What are porous bearings? What are the general requirement of bearing materials. 15

7. With the help of a diagram explain the working of a gas bearing. Under what conditions, the gas bearings are used? Discuss properties of gas for gas bearings. 15

8. What do you mean by elasto-hydrodynamic lubrication? Where this type of lubrication found in actual use? How the rolling contact bearings find their use in cams and gears. 15

**BT-6/J08**

**Tribology**

**Paper : ME-304E**

**Time : 3 Hours** **Maximum Marks : 100**

**Note:** Attempt any **FIVE** questions, taking at least **ONE** question from each unit.

## UNIT–I

1. (*a*) Explain tribological systems and their characteristic features. Explain one system in detail. 10

   (*b*) State and explain all the laws of friction and explain the working of cross-cylinder arrangement of measurement of friction with sketches. 10

2. (*a*) Explain different techniques of surface examination in detail. 10

   (*b*) What are the causes of friction? Explain adhesion theory. 10

## UNIT–II

3. (*a*) Define wear. Explain in detail adhesive wear. How do you measure the adhesive wear? 10

   (*b*) Explain fretting, corrosion and surface fatigue wear. 10

4. (*a*) What are the quantitative laws of wear? Derive the wear equation. 10

   (*b*) What is the mechanism of wear? Which type of wear finds application in industry? 10

## UNIT–III

5. (*a*) Explain elasto-hydrodynamic lubrication. 10

   (*b*) With the help of sketches, explain rolling contact bearings. 10

6. (*a*) Explain hydrodynamic lubrication with the help of neat sketches. 10

   (*b*) Give important properties of lubricants. 10

## UNIT–IV

7. (*a*) With the help of neat sketch explain hydrostatic bearing. 10

   (*b*) What are the important properties of bearing materials? Discuss in detail. 10

8. (*a*) Explain thrust bearing with a sketch. What are porous bearings? 10

   (*b*) Explain the working of gas bearings with the help of diagram. 10

## IMPORTANT QUESTIONS

**Important Questions according to Units occurred in Book:**
**Question and Answers generally set by the University Examiners:**
**For Answers see Index and Page No: according to topic**

1. Define Tribology and several disciplines in the field of Tribology.
2. Explain the term of wear and its classification.
3. Classify mechanism of wear.
4. Define wear coefficient and measurement of wear.
5. Write short notes of:
   (*i*) Quantitative law of wear
   (*ii*) Two body and three body Abrasive wear
   (*iii*) Wear Resistant materials.
6. What is friction and explain types of friction?
7. State law of friction and explain friction theories.
8. Explain rolling friction with coefficient of friction.
9. Illustrate stick slip phenomenon and angle of response.
10. Explain static and kinetic friction.
11. Explain friction measurement methods.
12. Explain lubrication, lubricants and function of lubrication.
13. Various properties of lubricants and types of the lubrication.
14. What are lubricants additives and classify them?
15. Write short note on:
    (*i*) Hydrodynamic lubrication
    (*ii*) Hydrostatic lubrication
    (*iii*) Boundary lubrication
    (*iv*) Elastohydrodynamic lubrication
    (*v*) Cam and gears.
16. Define bearings, Design of Bearing, Bearing Geometry.
17. Explain pressure distribution of journal bearing.
18. Reynolds equation for squeeze films and continuity of flow in column.
19. Explain rolling contact Bearing and Bearing materials.
20. Briefly explain tribological system and surface topography.
21. What are surface examination techniques?

# INDEX


## A

Abrasive wear 13
Adhesion/cohesion theory 52
Adhesion in friction 61
Adhesion wear 17
Aerostatic bearings 105
Angle of friction 49
Animal lubricants 75
Applied surface chemistry 7

## B

Bearing characteristics 110
Bearing geometry 94
Bearing materials 110, 111
Bearings 92
Benefits of metalworking fluids (coolants) 124
Boundary lubrication 83
Boron nitride 78

## C

Cam 91
Cavitation 97
Cavitation wear 22
Characterization of surface topography 117
Coefficient of friction 47
Continuity of flow in a column 102
Corrosive wear 24

## D

Design of bearings 92
Deterministic vs. Probabilistic (stochastic) 125
Diffusive wear 32

## E

Elastohydrodynamic lubrication 84
Energy of friction 49
Erosive wear 20

## F

Fatigue wear 27
Forces and stresses of rolling-contact bearings 109
Fretting wear 28
Friction 40
Friction measurement methods 57
Friction plot 56
Functions/objective of lubrication 76

## G

Gas lubricated bearings 105
Gear 85
Gear profile 89
Graphite 77
Grease lubrication 66

## H

Helical gears 86
Hydrodynamic or full fluid film lubrication 81
Hydrostatic bearings 103
Hydrostatic lubrication 84

## I

Impact wear 32
Internal gear 86

## J

Journal bearings 94

## K

Kinetic friction 49, 55

## L

Law of friction 42
Learning tribology 6
Lubricant additives 72
Lubricant impurities and contaminants 70
Lubrication 4, 63

## M

Machining tolerance 93
Magnetic particle inspection 122
Meaning of tribology 2

Measurement of wear and wear coefficient 12
Mechanisms of wear 11
Melting wear 31
Metal polymer adhesion 19
Mineral lubricants 74
Molybdenum disulfide 78

**O**

Oxidative wear 36
Oil lubrication 66

**P**

PAML model 12
Polimers blends 35
Porous bearings 99
Powder composites 36
Properties of lubricants 67
Purpose/functions of lubricants 64

**Q**

Quantitative expression for abrasive wear 10
Quantitative law of adhesive wear 18

**R**

Reynolds equation 100
Reynolds equation for squeeze films 98
Rolling-contact bearings 108
Rolling friction 44, 53

**S**

Solid lubricants/dry lubrication 76
Sommerfeld number 96
Spur gear 86
Static friction 48, 54
Stick-slip phenomenon 51
Surface examination techniques 119
Surface topography 115
Surface roughness 118

**T**

Terminology in wear 11
Theories of friction 43
Thrust bearings 106
Tribological systems 113
Tribology 1
Tribology of polymer composites 34
Tribology of polymers 33
'Tribos' 2

**W**

Wear 5, 9
Wear of ceramic materials 33
Weight loss techniques 39
Work of friction 50
Writing and sliding 59